HEAR *my* VOICE

An Old World Approach to Herding

by Lynnette Rau Milleville

Rowe Publishing

Copyright © 2015 by Lynnette Rau Milleville.
All rights reserved.

ISBN 13: 978-1-939054-36-4
ISBN 10: 1-939054-36-2

Back cover photo by Jeanine Dell'Orfano.

No portion of this work may be used or reproduced in any manner whatsoever without written permission of the copyright holder and/or publisher.

The information shared in this book may or may not be right for your situation. If you choose to use these methods, you will hold the author, publisher, and their affiliates harmless.

1 3 5 7 9 8 6 4 2

Printed in the United States of America
Published by

Rowe Publishing

www.rowepub.com
Stockton, Kansas

Dedication

In Memory of Robbie Kaman

In the mid 1980s, I met Robbie Kaman at a herding instinct test in Connecticut. Robbie and her husband Charlie Kaman were co-founders of the Fidelco Guide Dog Foundation. Fidelco uses German Shepherds as guide dogs for the blind. The foundation for their breeding stock came from working sheep dogs in Germany. Many of these dogs came from Karl Fuller's Kirschental Kennels in Stettbach, Germany. As my relationship with Robbie grew, she opened my eyes to the beauty of the German Shepherd and the work of the tending dogs. She made numerous dogs available to me to train and provided opportunities for me to study herding in Germany. Over the years Robbie encouraged me to teach others about tending. Writing this book is just one way I can continue to encourage others to open their eyes to different agricultural practices and set aside preconceived ideas about herding dogs from central Europe.

Acknowledgments

Hear My Voice has for 20 years, been a work in progress. During this time, many people have offered their expertise: Jo Steele gave me computer technical support. Nancy Spector started me on my writing journey. Lincoln Keiser (Linc) picked up where Nancy left off by gently taking me by the hand and teaching me how to write. This book would not exist if it were not for the many hours we spent together. Thank you so much. Along the way, Dorit Vanderwilden drew the basic figures for the illustrations. The structure figures were illustrated by Betsy Levine and Ben Kann. Pat Gauthier completed the final edit of the book. Without your help and the countless others who spent hours reading chapter drafts and offering suggestions, this dream would not have become a reality. Finally, thanks to my husband Tim, who behind the scenes encouraged me, believed in me and kept me grounded.

Contents

Introduction 1

One
The Development and Use of Tending Dogs in European Agriculture 6

Two
Sheep 16

Three
Shepherding 37

Four
Wild Canines 56

Five
Good Herding Dogs 77

Six
Learning 107

Seven
Basic Obedience 121

Eight
Pre-Livestock Foundations 137

Nine
Basic On-Stock Training 151

Ten
Advanced On-Stock Training 177

Eleven
Competitions 199

Twelve
Tending Applications 234

Appendix I
Sheep Breeds 249

Appendix II
Plants Poisonous to Sheep 253

Appendix III
Dog Breeds. 272

Appendix IV
Breed Contacts. 278

Appendix V
Puppy Aptitude Test Sheet 279

Bibliography 281

Videography 288

Introduction

I tell you the truth, the man who does not enter the sheep pen by the gate, but climbs in by some other way, is a thief and a robber. The man who enters by the gate is the shepherd of his sheep. The watchman opens the gate for him, and the sheep listen to his voice. He calls his own sheep by name and leads them out. When he has brought out all his own, he goes on ahead of them, and his sheep follow him because they know his voice. But they will never follow a stranger; in fact, they will run away from him because they do not recognize a stranger's voice (*John.* 10. 1-5; *New International Version Bible*).

The shepherding abilities of all breeds of herding dogs stem from the same source, the wolf's pack-hunting instinct. Through selective breeding, mankind has modified and molded this instinct in different ways to meet the specific needs of place and time. Over the centuries, three general herding styles have emerged: fetching, as typified by the Border Collie of Great Britain; driving, as represented by the Huntaway of New Zealand; and tending, as typified by the German Shepherd Dog. This does not mean, of course, that Border Collies only fetch, Huntaways only drive, or German Shepherds only tend. Some German Shepherds, for example, make excellent all-around farm dogs with the ability to fetch, drive and tend. It does mean that the best German Shepherd will not show the fetching ability of the best Border Collie. Similarly, a skillful trainer can teach some Border Collies to bark at stock. But their "noise" (to use the New Zealand term) will not have the power of the best Huntaway, a dog bred to use the force of its voice to drive sheep.

This is not to say that any particular herding breed is superior to another, as some have claimed. It does mean, however, that we must know our dogs' herding heritage so we can pick the breed that best meets our specific needs. For example, on a small farm in the hills of New England or on the rolling countryside of the Midwest, a loose-eyed,[1] all-around farm dog might be the best choice, while a fetching dog might be better on a large sheep ranch in the Rocky Mountains. In the cut blocks and municipal grazing areas of Western Canada, a tending dog might best meet our needs. Only by carefully matching a breed's inherent strengths and weaknesses with the tasks required, can we make the best choice. We will enjoy only limited success if, on the contrary, we require our herding dogs to perform tasks to which they are not ideally suited.

Similarly, we should design our training techniques to maximize our dog's strengths and minimize its limitations. Further, we must keep in mind that strengths and limitations can be defined only relative to the herding tasks at hand. This means that no single training technique or program is best for all breeds in all situations.

Some successful trainers argue that the techniques developed to train Border Collies work best for all breeds with the exception of some very specialized driving breeds like the Huntaway. I disagree. We will train our dog more efficiently and in less time if we develop training programs designed to enhance our dog's unique talents.

The same principles apply to dogs trained specifically for herding trials. In Canada, Europe and the United States, most of the dogs working livestock are sport dogs, trained for trialing. Relatively few work on ranches or farms any longer. If you wish to train your dog for the sport of trialing, you will enjoy more success if you choose the style of herding and course for competition that best tests the special capabilities of your dog and if you choose a training program that teaches your dog to work the course you have chosen. Specifically, if you own a Border Collie, use a training regimen that prepares you and your dog for the International Sheep Dog Society (ISDS) course. If your dog is a German Shepherd, train to compete on the Verein für Deutsche Schäferhunde (SV) course. If your dog has all-around herding talent, particularly with driving livestock, train for the Australian Shepherd Club of America (ASCA) course.

This is a controversial position; many disagree with it. I urge you to read widely to become acquainted with other points of view, so you can better choose the approach that best suits you. Most sheep herding programs are based on relationships of fear: the sheep's fear of shepherds and their dogs. In contrast, this book is about building trust. The sheep learn to follow the shepherd, not because they must, but because they want to. Over the years, I have trained a variety of working dogs, fetching dogs, all-around herding dogs and tending dogs. I have tried different techniques and programs. The kind of specialized approach detailed in the following pages has worked best for me, as well as for my students. This book is about training dogs to tend sheep in the European style.

---

1   A strong-eyed dog shows intense concentration, a stalking approach and a natural tendency to pause or crouch. A medium-eyed dog shows intense concentration, but freer movement, without stalking or pausing. In comparison, a loose-eyed dog has good concentration, but lacks the intensity of the strong or medium-eyed dog. The loose-eyed dog may glance at its surroundings, but this does not signify a lack of attention. The loose-eyed worker is often as effective as one that shows more eye (The American Working Collie Association n.d.:6).

I will never forget my introduction to herding. I was awestruck with the magic of a Border Collie named Roy. The grace and fluidity of his movements, his complete focus on the sheep and his instant obedience to commands was amazing to behold. For most of us who have watched these dogs on TV, the Border Collie is the sheep dog *par excellence*. I have been involved in the breeding, training and trialing of Border Collies for many years and know firsthand the amazing feats that distinguish this remarkably capable dog. Thus, I can understand why some find other herding breeds wanting when compared to the Border Collie.

Yet, the Border Collie, as all herding breeds (indeed, all working breeds) as a product of human technology, was born into the context of specific social, cultural, political and economic developments at a particular historical place in time. Because this book is not about Border Collie training per se, it is not necessary to detail the history of the breed here. The Border Collie is a child of the Industrial Revolution—bred in Great Britain to herd in a particular way to meet the needs of the time.

Specifically, the job of the Border Collie was to gather flocks of sheep pastured in mountainous areas for long periods of time. Natural predators were few in 19th century England and sheep could generally fend for themselves. They needed neither the close supervision of a shepherd, nor the protection of a dog. The close, trusting relationship between shepherd and sheep that remained important in other areas of Europe became obsolete in the border regions of Scotland and England. Before the British Parliament passed the Acts of Enclosure, shepherds were forced to remain with their flocks for long periods of time. After enclosure, this became unnecessary. Consequently, shepherds no longer built close, trusting relationships with their sheep: no longer witnessed the birth and death of each member of their flock. In this new shepherd-sheep relationship, the dog became the enforcer. It ran far and wide, as its wolf ancestors, forcing the sheep to flock together and then drove them to the shepherd's feet through its ability to intimidate.

In contrast, the shepherds of central Europe continued to use their sheep dogs as they had for centuries. To this day, Shepherds in Germany work closely with their flocks to gain their trust (although, on occasion, their dogs must punish sheep for disobedience). In this tending relationship, the sheep must hear their shepherd's voice. The dog in this herding equation does not intimidate the sheep by barking (as does the New Zealand Huntaway) or coerce the sheep with its "eye" (the intense stare of the Border Collie). German Shepherd dogs form "living fences" to protect sheep from predators and crops from sheep. National herding competitions in Germany still feature this inherent ability of the working German Shepherd and other native herding dogs.

Without understanding the interdependent relationship between shepherds and their sheep, attempts to train dogs to tend sheep in the European style will enjoy only limited success. In the tending equation, the shepherd is the leader and the dog his helpmate, while the sheep look to both for food, shelter and protection. To grasp the dynamics of this equation, we must explore not only the perspective of the shepherd and his dog; we must understand the sheep's point of view, as well. This book differs from others about herding stock with dogs in content and in order of presentation.

The book is divided into twelve chapters dealing with topics such as the history and development of tending dogs, shepherding, choosing a herding dog, learning, obedience, and stock training. Undoubtedly, many readers will want to skip the early chapters and get right to the fun stuff—training dogs to tend sheep. I certainly enjoy that the most. But the order is important, for each chapter is a bridge to the next. Moreover, many of the techniques described in this book are specific to the training of tending dogs, and some are controversial. To understand the reasons why these techniques work, and to make them work better for you, it's necessary to put them into context.

The book begins with the dogs themselves and views the ways in which they developed in particular social and ecological contexts. It then moves on to a discussion of sheep. In order to move sheep efficiently, (or any livestock, for that matter) you must first know something about both their behavior and their nature. For example, you must know what they fear, how far and from what direction they can see, how they move and what they eat.

To become good at herding, you must become a good shepherd. You cannot do that until you know something about the art of shepherding itself. As we shall see in Chapter 3, this knowledge comes from a surprising source, the Bible's 23rd Psalm. Through reading and analyzing it carefully, we can learn the basics of becoming a good shepherd; it is a flock care checklist.

Once you understand good shepherding, you must learn herding from your dog's perspective. Dogs are descended from wolves; canine herding is rooted in the wolf's powerful instinct to hunt and kill prey. Although human beings have modified and molded the dog's hunting behavior and suppressed the instinct to kill, the herding instinct remains extremely powerful. Because the line between herding and hunting-killing is so thin, once the instinct is awakened, it must be carefully channeled and controlled. You must decide whether or not you want this, for once awakened, the predator/herding instinct can become a driving force in your dog's behavior.

Now you are ready to choose your dog. In order to make the best selection possible, you need some way to evaluate potential choices. Chapter 5 discusses how best to approach this seemingly daunting task. It includes guidance about how to test for the traits you desire in a dog and information on the techniques of structural characteristics assessment.

The way dogs learn to obey commands is the subject of Chapter 7. Here, we look at the principles of operant conditioning, for they provide the foundation on which learning rests. However, the principles of operant conditioning must be put into practice.

Without achieving a balance between your control of the dog and the dog's control of the stock, you will be unable to move the sheep from place to place or to keep them where you want them. You will certainly never get them back into the barn after they have been in your neighbor's garden. This chapter deals with basic obedience, which is the foundation for herding training.

We are at last, ready for the part you have been waiting for, how to train dogs to tend sheep. We begin with a short chapter on pre-livestock foundations. In Chapter 8, you will learn, among other things, how to build a training area, how to teach your dog to jump and how to condition your flock.

Now we move to actual work on stock. The next two chapters cover the main elements of tending—on-off lead training, moving your flock within a given boundary area, gripping, and passive and active control of your flock. In addition, you will learn about penning and road and bridge work, including how to herd sheep in traffic.

For those who wish to pursue stock tending as a sport, Chapter 11 discusses trailing and judging. It describes the levels of difficulty in tending competitions, the skills needed by shepherds and their dogs, and the basic guidelines for performance herding. The smart handler is also a judge, and in this chapter, you also learn how judges evaluate you and your dog's performance.

We finally turn to tending applications—how to use the tending method in real life situations. In particular, Chapter 12 discusses tending for American farm operations. In it, we learn to think of sheep as both tools to use and as livestock to raise.

So, if you are a farmer who wants to use his dog as a living fence while rotationally grazing his stock, or a person who simply wants the fun of working with his or her dog in competitions, this book is written for you. This book also contains helpful information for instructors and judges.

One

The Development and Use of Tending Dogs in European Agriculture

If you own a tending dog, you have acquired an animal with a long history in European agriculture. Its ancestors were utilized by shepherds, drovers, farmers, butchers and herdsmen. Each of these professions required a different type of dog with different skills—thus, the many breeds of dogs with herding ancestry.

For centuries, European herding dogs played a key role in agriculture and still do in many European farming communities. In order to understand how these dogs developed, we must first understand the agricultural system in which they worked. According to David Grigg, world agricultural systems fall into the following types: shifting agriculture, wet rice cultivation, Mediterranean agriculture, pastoral nomadism, ranching, dairying, plantation agriculture, large scale grain production and mixed farming (Grigg 1974:3). Of these, Mediterranean and mixed farming are most common in Europe.

The Mediterranean agricultural system is found in the areas surrounding the Mediterranean Sea. These areas are typified by mild, wet winters and hot dry summers. A small portion of the land is arable and the rest is unimproved pasture. Summer droughts in the areas surrounding the Mediterranean Sea affect the quality of pasture. Therefore, few beef and dairy cattle are raised in this region. Sheep and goats are the livestock of choice because they can survive on poor quality grazing. "Thus, a typical village community would grow wheat and barley in the plain, graze sheep and goats on the stubble, grow olives and grapes on the lower hills with patches of irrigated vegetables around the village. In the summer the flocks would be driven to the higher pastures of the mountains to return to the plains when autumn rains came" (Grigg 1974:123).

Throughout the Mediterranean region the seasonal migrations between summer and winter pastures (called transhumance) reached their peak between the 15th and 19th centuries. By the late 1800s, however, the development of railways made walking livestock to and from seasonal pastures unnecessary. At the same time, never before used grazing land opened up for use. Today, long distance transhumance is found in parts of Greece, Italy, former Yugoslavia, Spain and North Africa, while shepherds make shorter treks consisting of three-to-four day journeys in southern France.

Farming in north central Europe was of the mixed variety. This system is generally associated with densely populated and often industrialized areas. Farms are owned and managed by families who provide 70-85% of the necessary labor needed to operate them. In Germany, farms are approximately 25 to 124 acres in size. Farms are larger in northern France and Great Britain. In Sweden, Denmark, and Great Britain, farmland is generally consolidated, while in other areas of Europe it is more fragmented (Grigg 1974:154).

> "(T)he bulk of farmers in western Europe lived not in isolated dwellings or hamlets but in villages, and walked out each day to work in the arable fields surrounding the villages. There was a marked zonation of land use in the open fields. Near the village were closes[2] of grass for livestock, and gardens of intensive cultivation. Surrounding this were the great open arable fields, which were divided into long, narrow strips: each farmer's strips were intermingled with other strips rather than in contiguous blocks. Surrounding the arable fields was common grazing land and woodland" (Grigg 1982:141).

Farmers typically grow three types of crops: cereal, root and grass.[3] Grass accounts for approximately 1/5 to 3/4 of the total arable land.

In Western Europe, 3/4 of the arable land grows food for animals and most farms are self sufficient in this respect. In the Netherlands, 67% of feed available to farm animals takes the form of grass. Only during the winter months is livestock fed forage crops and concentrates. Thus, most livestock in Europe harvest forage themselves.

Throughout Europe, very little fencing separates crops and animals. Consequently, dogs are often used in place of fences to protect crops. In certain areas, however, this is changing, since the number of fenced pastures is increasing.

Early Mixed Farming

Between the 14th and 20th centuries, three major trends changed European mixed farming agriculture. These trends caused shepherds to refine and develop the abilities of their herding dogs, changing their characteristics in the process. The first trend

---

2 Land adjoining a dwelling enclosed by a fence
3 Wheat, rye, and barley are grown for animal feed along with sugar beets, turnips, mangolds, potatoes, and swedes.

involved changes in how and what crops were grown. During the 14th century in Flanders, Zeeland and Friesland, farmers began to grow turnips on land previously left fallow. Turnips were a good source of winter feed for cattle. In turn, the cattle provided fertilizer for the fields. Later, farmers began to use clover in the crop rotation system.[4] When fields were planted with clover, the soil replenished its lost nitrogen. In addition, the fields provided fodder for the animals.

By the early 1800s, farmers had established a four-year rotation system. They first planted turnips, then barley or wheat, then clover, and finally, a cereal crop. In the late 1800s, farmers integrated livestock and crop production and mixed farming became the norm in the region.

The second trend involved a change in the type and numbers of livestock raised. Between the 14th and 18th centuries, sheep production increased as a market economy replaced subsistence agriculture. In the process, sheep breeds were specialized to improve meat, wool and milk production.

With the Industrial Revolution, European agriculture changed even further. As industrialization increased, the human population exploded. With this explosion in population, land use patterns necessarily changed. Agricultural land holdings become more fragmented. As a result, many farmers had to become even more concerned with keeping their animals out of their neighbors' property because they could not afford the fines levied upon livestock trespass.

At the beginning of the 14th century, farmers used dogs mostly as guardians to protect livestock from predators. As the new system of crop rotation spread and the Industrial Revolution fragmented agricultural land holdings, farmers found it necessary to develop dogs that could protect crops from livestock. As more crops were introduced, the problem intensified, since animals often grazed in between the various crops. A farmer's dog had to pay close attention to its charges as they traveled on roads and grazed in pastures.

The third trend in European agriculture involved changes in transportation systems. By 1850, Britain had well developed railroads, followed by France and Germany in the 1880s. As a result, many farmers no longer needed to walk livestock to market centers (Grigg 1974: 167-168). This reduced the need for droving dogs and not surprisingly, their numbers began to decrease significantly.

After World War II, modern divided highways built to accommodate high-speed traffic began to crisscross Europe. As a result, the need for herding dogs changed even more. Dogs were no longer needed to escort the livestock to and from winter and summer pastures, for now, they could be transported by truck and/or train.

As long as fencing was impractical (because it was either too expensive or the terrain was too rough) and agricultural holdings were fragmented, the need for tending

---

4   The system consisted of two years cereal, one year fallow, and three-to-six years grass.

dogs remained. In some of the former Eastern Block countries, privatization of agricultural holdings, coupled with severe underdevelopment, have created an even greater need for the special skills and abilities of tending dogs. In Bulgaria, for example, tending dogs are important in dairy, meat and wool production.

Development of Tending Dogs

The development of tending dogs reflected increasing demands on herdsmen as agriculture intensified and diversified. When populations increased, open space diminished. Consequently, herdsmen and shepherds were under considerable pressure to keep their flocks under control while moving them to and from grazing areas.

The need for specific tasks or services prompted the quest for instinct development. Shepherds began to carefully select those specific herding abilities that conformed to the new kind of work required of their dogs. Many times we read that working dogs required no special training—perhaps because of this selection process. Farmers and shepherds severely culled their dogs for lack of desired herding abilities. A farmer could not afford to keep dogs that could not earn their keep. In many litters, only one dog was kept; the others were destroyed.

The selection of herding traits was not accomplished by using a one-time test. It started with an assessment of pups that began when they were still young and continued for many months. As time passed, dogs that could not perform were culled. Dogs were not forced to work. On the contrary, learning to herd happened naturally through daily exposure to livestock. In this respect, John Holmes, a noted author and trainer, made the following observation:

> "… It is of the utmost importance to remember that, when a young dog starts to run, he does so instinctively… The young dog herds, not so much because he wants to herd, as because he cannot help it any more than when he was born…" (Holmes 28).

Today, European farmers and shepherds use several breeds of dogs during the course of a day. For example, a shepherd in France may use a Briard to tend his flock during the day and a Pyrenean Mountain dog to guard it at night. Each breed is an expert at its given job. For it was selectively bred over many generations, its instincts sharpened in the process.

Most herding breeds reached their highest use in the late 1800s. As railroads began transporting livestock to market, the use of herding dogs diminished. Some breeds survived by performing other services, e.g., protection and family companionship. Still, some need for working dogs remained and the development of herding skills through selective breeding continued. As a result, the herding abilities of some contemporary breeds remain high.

Transhumance

As I mentioned before, transhumance is the migratory movement of livestock between grazing areas because of climate and vegetation restrictions. It is a way of life for many shepherds and their families throughout the world. In Europe, transhumance is primarily vertical, i.e., flocks move directly between lowland and highland pastures in response to seasonal changes. There are two eco-systems in Europe which encourage vertical transhumance. The first is the area close to the Mediterranean Sea while the other includes the high mountain areas of Western and Central Europe.

Mediterranean Transhumance

Rainy conditions during the winter months allow for winter grazing along the mild Mediterranean coast. Hot, desert-like conditions burn vegetation during the summer months. Flocks must be taken to graze in the cool upland pastures during the summer. As winter approaches, however, the upland pastures become unbearably cold and the flocks must return to the coast. Thus, vertical transhumance in the Mediterranean region is a function of climatic variation in summer and winter pastures.

In Mediterranean areas grazing is generally confined to non-arable land. The vegetation that grows there is more suitable to sheep than cattle. Moreover, these pastures are generally found at higher altitudes. Lowland farmers, however, welcome animals in the winter to fertilize their fields (Davies 1941:157). In doing, so they replenish soil, preparing it for spring planting and reinforcing the pattern of vertical transhumance in the process (Grigg 1974: 123-125).

Alpine Transhumance

The seasonal movement of stock in high Alpine mountain areas is more closely related to the pattern of crop cultivation than to climatic variation (Davies 1941:165). Farmers must use lowland pastures to raise hay in the summer, so that their animals have feed in the winter. Thus, the animals must vacate the lowland pastures during the summer growing season. Here, the high variation in temperature between highland and lowland pasture areas typical of Mediterranean areas does not exist.

The pastures in this ecosystem are of two types: high mountain above the tree line that is not free of snow until May or June of each year, and forest comprised of woods and some grass meadows. En route to the high mountain pastures, the herds graze in the forest. Consequently, one third of an animal's diet consists of brush found in these forested areas. Not only is this forage especially good for lactating cows, but it also produces excellent tasting butter, as well as the nutty flavor of Gruyere cheese. During the autumn, animals return along the same route.

When it is time to descend the mountain slopes, the cattle return first because sheep and goats are more resistant to frost. This also allows the sheep and goats to clean the pastures already grazed by cattle. When the cows return to the village, they find additional grazing in the stubble of cultivated fields surrounding the villages. In the winter, animals graze in marshes, permanent winter pastures and on the stubble of cultivated fields. In many instances, shepherds must rent winter grazing for their flocks.

Because the distance between summer and winter grazing could be as great as 100-500 miles, families in some areas of Europe owned two dwellings, one in the mountains and the other on the plains. Consequently, sister villages often developed along the routes to and from summer and winter pastures (Evans 1940:172-180).

National borders did not always hinder the movement of livestock. Crossing borders was, and still is, common practice among some shepherds. The border dividing France and Spain snaking its way through the Pyrenean Mountains provides an interesting case in point. This border was first set by the Treaty of the Pyrenees as a part of the settlement ending the Thirty Years' War in the 17th Century. Significantly, the treaty did not specify the border in the central and western Pyrenees region. On the contrary, it left the exact border to arrangements made by shepherd communities, a practice dating from medieval times.

The primary purpose of these local arrangements was to ensure the free movement of livestock, people and goods, thereby maintaining peace among the shepherd communities. As a result, the valley communities of the region kept peaceful relations with one another despite the wars between France and Spain. For example, during Napoleon's Peninsular Campaign (1812) valley communities "... resisted participation and collaborated to keep the mountains peaceful" (Gómez-Ibáñes 1975:45).

In the mid 19th century, a commission was established to formalize the border. However, the resulting treaty conceded to pastoral traditions, allowing local communities the right to make grazing arrangements with their neighbors as they saw fit. Consequently, shepherds and cowherds continue unrestricted boundary crossings to this day (Gómez-Ibáñes 1975:50).

E. H. Carrier's book, *Water and Grass* describes the seasonal migration of flocks, herds and people in southern Europe. Throughout the book, there are similarities between the techniques and customs handed down from generation to generation and country to country. The following account was made in Italy in 1838:

> The sheep were divided into flocks of about 10,000 head. A certain number of them, generally belonging to the same proprietor, were under the immediate management of an agent or factor, always a native of the Abruzzi. This factor accompanied the flocks mounted, and was armed with a musket against the surprises and emergencies of the march. The sheep were distributed in files of twelve, and white sheep dogs, running at the front, centre, and rear of

each group, kept up the speed and gave protection where necessary. Behind the sheep travelled a small portion of black goats. A shepherd with his crook marched in front, closely followed by an old ram with a bell tied round its neck, whose sonorous notes guided the flocks along the appointed route. The Abruzzi shepherds, numbering some five thousand, summer and winter alike wore large sheepskin jackets and thick, strong shoes. In the wake of the flocks followed cows, mares, and beasts of burden laden with the nets and poles required for the evening folding. Cloth tents for shepherds, utensils for milking, including portable and jointed stools made out of the stem of the giant fennel, pots in which to boil the milk, completed the baggage. The journey between the summer and winter home might last as long as a month, and both flocks and guardians found it very fatiguing (Carrier 1932:50).

During treks between winter and summer pastures, the dog's work is similar to its daily tasks in the winter areas, except that the work is carried out on a much larger scale. Once the mountain grazing areas are reached, however, the dogs are hardly used. Why? Mountain is different from plains grass management. For example, in mixed farming situations, sheep are encouraged to graze close together in assigned areas until they harvest all the fodder. These areas are small and bordered by other crops, thus requiring both the shepherd and dog to be keenly observant.

Mountain grazing is different. There, sheep are encouraged to pick and choose as they go because:

"Sheep, when allowed to graze too closely, will tear out entire plants by the roots. When this is done by such large numbers of sheep, it can reach disastrous proportions. The destruction of the vegetation always results in the eroding away of the mountains during the rainy season. Rain water, no longer held back by vegetation, rushes down the slopes in torrents, causing flooding and the washing away of the soil. Here, the methods employed on the plains are not sufficient. This is a different situation altogether" (McLeroth 1982:163-164).

You might think because grazing areas in the mountains are larger than on the plains, dogs would have more work to do. This is not the case, however. Sheep are much easier to oversee in large areas. They tend to relax as they eat because there is less competition for food and more of it to choose from. Consequently, both shepherd and dog can relax as well.

Let's take a close look at a typical day for a European shepherd and his flock. It begins with the shepherd taking his animals from their overnight fold and leading them along roadways to their various grazing areas. At the end of the day, he returns them to a pen to spend the night. Noël Wanlen's 1946 doctoral thesis (translated from French for Diane McLeroth's book, *The Briard*) shows in some detail the work of the herding dogs in France. The following account is based on this work.

Road Work

The shepherd must walk slowly, at the same pace of his flock, keeping in mind the sheep at the end of the column that must run to keep up. "The sheep should be well-grouped and should not run, especially if it is hot..." for then they will become over heated (McLeroth 1982:156). Moreover, shepherds should keep their hungry flocks from moving too quickly in the morning while en route to pasture. The combination of fast movement with a change from sparse grazing (in an overnight pen) to lush, dew laden forage in the morning will cause sheep to bloat.[5]

Two dogs work the flock: a "man" dog and a dog that works at a distance. The "man" dog stays with the shepherd and works the same side of the flock as he, while the dog at a distance works the opposite side. Both dogs must patrol the flanks of the column to prevent the flock from lingering and grazing along the shoulders of the road.

Once the flock sets the pace, the shepherd can fall back to the rear of the column. From there, he can oversee the flock without having to walk backwards. If the flock's pace changes, he must return to the head of the column. If he does not do so, the flock will either run off without him or stall, spreading out and blocking the roadway.

The most difficult task when traveling along a roadway is to keep the flock moving in a straight line while vehicles pass by.

> "In order to do this, the shepherd calls in the dog at a distance, who would usually be on the right side of the flock. Then, he goes to the head of the column and presses to the right. The first animals follow him, then the entire flock. At the same time that he executes this maneuver, the side dog, now on the left, at the shepherd's command, also presses over to the right side of the road. He keeps them there while "balancing", that is to say, while going and coming from one end of the column to the other, to bring in the rebels who try to escape" (McLeroth 1982:157).

Once the vehicles have passed, the shepherd returns the flock to the center of the road. Each dog then returns to its original duties, i.e., balancing the flock on the road.

"(M)ost roads are clearly defined by ditches and embankments on one side or the other. Normally the sheep do not leap over these obstacles" (McLeroth 1982:157).

Sometimes roads are not so clearly separated from the land that borders them. Only a sliver of grass may separate the road from a bordering wheat field.

---

5   Bloat is an excessive accumulation of gas in the rumen. Symptoms include the enlargement of the rumen on the sheep's left flank. Generally, sheep will have difficulty breathing and grind their teeth in pain because of the gas build-up. If not treated, the condition can be fatal in as little as two hours (Simmons 1989:76).

"The dogs then, are hard-pressed to see that the sheep are not enticed to go parading into the fields" (McLeroth 1982:157).

When approaching a narrow road, the shepherd must draw out the length of the column. If he does not, the flock will slow its pace and in some cases, stop. To do this, the shepherd returns to the head of the column and accelerates the pace. The lead sheep soon follow at the quickened pace, thus stringing out the column.

"In the meantime, the flanks of the flock are carefully watched by the dogs. The dogs accomplish rough work, demanding their presence along the entire column. It is a very beautiful spectacle, to see the faithful guardians coming and going constantly along the sides of the flock, preventing the sheep from going beyond the edges of the road" (McLeroth 1982:157).

When approaching a sharp turn, a shepherd must be careful that his flock does not trample the crops along the roadway. Sheep have a tendency to cut corners, trampling the crops along the side of the road in the process. Generally, the first sheep will not cut the corner but the rest will. If the shepherd is not observant, the entire flock will walk into the adjoining field, trampling all in its path. To avert this, a shepherd places himself and one of his dogs at the dangerous corner, while the other dog continues to patrol the sides of the flock, keeping them on the road.

Grazing

Unattended sheep tend to roam about freely. When first arriving in a new pasture, they wander in every direction. They spend their time walking about trampling the very grass they should eat, thus damaging the pasture instead of improving it. A shepherd should keep the flock together, only allowing them to graze in the area he assigns to them before moving them forward. If the animals cannot wander about, they will find nothing to eat but the grass available. Thus, they will eat the more palatable grass first and the rest later. Only then should the shepherd allow the flock to continue their forward grazing.

The shepherd should divide the grazing area generally into squares and graze each one in succession. The dogs must patrol the boundaries of each square in a way that does not allow sheep to cross into another square. When working two dogs, one will work the side with the shepherd while the other works the opposite side. The shepherd relies on this dog to work on his own.

"Without being told, he understands how to prevent the sheep from crossing over the limits of the pasture, which have been entrusted to him. If the sheep are numerous, or particularly quick, it may be necessary to use two "side dogs", each watching one side of the meadow. Supervision of the two other sides is ensured by the shepherd himself, assisted by the dog "on the man".

This dog stays near his master, not moving about, but rather lying down. His role consists in returning an escaping sheep to the flock, within the limits of the square, responding to a simple gesture, whistle, or calling of his name" (McLeroth 1982:160).

The "man dog", the dog that works on the same side as the shepherd, does not need to show the same initiative as the dog that works the opposite side. It is here that the young dog begins his training on a chain.

"When it is time to return to the fold, the shepherd commands his dogs with voice and gesture to patrol the sides of the flock. As the dogs balance the flock, the sheep form a column and are led home by the shepherd. Again the shepherd must be careful not to let the sheep move at a rapid pace. He must hold them back, keeping their pace slow" (McLeroth 1982:156-161).

Today, many people are interested in the different breeds of herding dogs and intend to bring these dogs back into useful service. Some do this by looking at history, concentrating on recreating the characteristics that typified these breeds in the past. Conversely, others choose to develop in their dogs herding characteristics useful under today's conditions. Regardless of the approach taken, training for both trial and farm work will be easier if necessary herding tasks are first carefully identified and then seriously considered in setting the goals of breeding programs.

Two

Sheep

Mary had a little lamb,
Its fleece was white as snow,
And everywhere that Mary went,
The lamb was sure to go.

Domestication

Early man began domesticating sheep in southwest Asia about 9000 BC. Imprinting was the major catalyst in the process (Ryder 1983:3, 1984:66). Lambs reared by people from birth usually imprint on those people. They bond and react to human beings as they would their ovine mothers.

Mary Sawyer's little lamb exemplifies this process in action. Early in the spring of 1814, Five year old Mary adopted a weak, newborn lamb which she bottle fed and nursed to health. Mary played with the lamb (named "Snow") throughout the summer and was rarely seen without him. When summer ended, Mary attended school and her long romps with Snow ended. Each morning before leaving for school, Mary faithfully visited Snow and each time she did, the lamb tried to follow her. Mary always put the lamb in its pen before she left. One day, after Mary left for school, her brother opened the gate to Snow's pen. He wanted to see if Snow would follow Mary the half mile from the farm to school. Of course, he did (O'Donnell 1988: 16-17). The story of Mary's little lamb provides a picture of how imprinting works. Snow accepted Mary as his mother in the same way wild sheep 11,000 - 12,000 years ago bonded to nomad women, thus, beginning the domestication process (Ryder 1983:26).

The story of Snow, however, does not illustrate the consequences of the process, for over the years, domestication changed many of the sheep's physical characteristics. Wild sheep range in size from 80 to 450 lbs. Their coats are comprised of hairy outer and woolly under coats. Color combinations of black, brown, gray and white provide excellent camouflage (the white hair is found on their bellies). In addition, wild sheep have short, thin tails, long horns, slender long legs and relatively small hearts, eye sockets and brains (Ryder 1984:69).

You can still find semi-domesticated, primitive sheep called Soay on the Island of St. Kilda in Scotland. These sheep bear a strong resemblance to their wild ancestors in several ways:

1. Soay sheep are small, weighing only about 55 pounds;
2. They have short tails;
3. Their fleece is multi-colored with a white belly;
4. They shed their coat in the spring; and
5. Both genders have horns

Our ancestors wanted small, hornless sheep with a single colored, wool fleece. They did not want a multi-colored coat of hair fibers. The reasons for this were:

1. Wool fibers hold dye better than hair fibers;
2. Light colored sheep are easier to locate than camouflaged sheep; and
3. Wool fleeces do not molt in the spring

These characteristics enabled shepherds to gather wool by shearing the sheep, a process easier than having to look for small clumps of wool shed on the ground (Henson 1991:9).

Human beings bred sheep for different reasons. Some were raised for food and fiber, while others were pets. Generally, sheep were raised for fiber, pelts, meat and milk. Today, 5.1% are used for meat, 0.6% for dairy, 10.9% for wool, 1.3% for meat and dairy, 53.7% for meat and wool, 3.9% for dairy and wool, 24.5% for meat, dairy and wool (Lynch, Hinch and Adams 1992:3).

Wild sheep still graze in many of the world's mountainous and high plains regions. Several species exist, including Bighorn found in the Rocky Mountains of North America, Argali in Central Asia, Ural in South-west Asia, and Mouflon, found in Western Asia and Europe (Ryder 1984:64). Archaeological experts believe that mountain sheep were domesticated first (Ryder 1984:66) because they were easier to catch than the faster plains animals (Henson 1986:9). Historians believe Ural sheep were the first to be domesticated.

Photograph 2.1 Rocky Mountain Bighorn Sheep, Banff National Park, Alberta, Canada.

Later, the European breeds were derived from the Mouflon, while the Argali contributed to the development of the Asiatic breeds (Ryder 1984:65, Lynch, Hinch and Adams 1992:2-3).

More than 2,000 breeds of domestic sheep exist today. Of these, 311 are "main" or dominant breeds. The other "minor breeds" were derived from them. The Australian Merino, for example, is a dominant breed while the Polled Merino is a minor breed, derived from it. Readers can find a breed list in *A World Dictionary of Breeds, Types and Varieties of Livestock* by Ian L. Mason. A short list is included in Appendix I.

Use

As you can see, sheep producers have many different types of sheep from which to choose. For instance, you may live in a hot region and want a breed that is heat resistant (Barbados). Alternatively, you may live in an area that requires sheep to be excellent foragers (North Country Cheviot). *Raising Sheep the Modern Way* by Paula Simmons and *The Sheepman's Production Handbook* by the Sheep Industry Development Program are excellent sources for more detailed information on the characteristics of various breeds.

Photograph 2.2 **Merino sheep in Germany**

Management

M.L. Ryder's book *Sheep and Man* documents the close connection between sheep and human beings. Over time, this connection evolved into two management styles: controlled and free grazing.[6] You can free graze sheep in large fenced or unfenced areas with little human contact, which leads them to become semi-wild. This method requires Border Collies or similar breeds of herding dogs to help gather and move the flock. Alternatively, you can control graze your flock with or without fencing. This requires both you and your dog to oversee your flock during the day and to pen the sheep at night. In the case of controlled grazing with fences, i.e. rotational grazing, fencing is moved routinely. In either case, sheep become accustomed to the presence of people, so much so that they follow them, knowing they will be led to greener pastures.

The relationship between human beings and sheep directly influenced the methods of grazing. As a result, two general systems of grazing evolved, transhumance and the confinement of sheep to fenced grazing areas. Which works better?

Scientists studying the effects of both management systems found that each one has advantages and disadvantages. Free grazing sheep within a confined area allows them to graze at their own pace. They are allowed to pick and choose their forage. Because they are left to fend for themselves against predators, their defense drive is heightened. Since the sheep are confined, their feces are easy to collect. Because the sheep are controlled by fencing, the shepherd can attend to other farming tasks, e.g., repairing fences and producing feed.

Unfortunately, free grazing also has a down side. Sheep overgraze the more palatable grasses first, leaving other areas under-grazed. Plants in the under-grazed areas become over-mature, tough, dry, and unpalatable" (Kruesi 1985:31). As choice fodder is overgrazed, leaf regrowth is diminished, thus, reducing the plants' ability to produce their own food. This also causes plants to draw on root food reserves (University of Vermont 1984). In turn, sheep need more land to graze because they trample their fodder while looking for the best grass available.

Sheep are often lost when free grazing. They may stray from the rest of the flock using undetected weaknesses in the fencing and wander miles before being missed. Others will cast on their backs (lay down with feet up) and are unable to get up. They may become entangled in brush which renders them easy targets for predators. Others become sick and die because they are not treated in a timely manner.

When sheep are confined to the same area, manure builds over time. This causes over fertilization of the soil, which in turn kills plant life. Parasite infestation also becomes a problem, requiring frequent use of expensive worming medications.

---

6   When sheep free graze, they are allowed to roam at large at their own pace, approximately 5.9 km/day (El Aich and Rittenhouse 1988: 279-290).

On the other hand, transhumant grazing requires flocks to be moved and overseen continually. Shepherds are always nearby to protect their charges from predators, poisonous plants and each other. Because shepherds are with their flocks 24 hours a day, they are involved with every aspect of the sheep's lives, from birth to death. This intimacy makes it easier for shepherds to care for their animals. Often, they can observe individuals without disturbing them as they graze or lie down. This awareness can alert shepherds to health problems before they become obvious, i.e., foot rot or pneumonia.

While the flock benefits from transhumance, the land does as well. Soil erosion is of small concern because flocks are not allowed to overgraze an area. At the same time, flocks systematically graze the areas, leaving them evenly cut.

Because shepherds go into the grazing area before the flock, they know the types of vegetation available and can maneuver the flock away from poisonous plants. If shepherds notice a particular area is potentially dangerous, they can protect their flocks by using dogs as living fences to keep their animals away from the area.

As with free grazing, transhumance has a down side. Because shepherds tend their flocks 24 hours a day, 7 days a week, their participation in other farming activities is limited. Moreover, they must scout areas in which to graze their flocks in advance. When shepherds are sick or wish to leave for other reasons, they must hire someone to tend their flocks.

Substitute shepherds tend sheep for pay and are sometimes not as careful as self-employed shepherds who care for their flocks because they have a personal interest in them. Self-employed shepherds are committed to their charges and willing to put themselves out for their flocks, sometimes risking their own lives in the process.

An overlooked weakness of transhumant grazing is the fact that sheep do not spread out and graze naturally or walk at their normal pace. Often, sheep are hurried from place to place by inexperienced shepherds, causing them to lose weight. Even when they are allowed to graze they are sometimes harassed by poorly trained dogs, causing some sheep not to eat at all.

A study conducted in the Middle Atlas Region of Morocco provides a case in point: The objective of the study was "…to investigate the impact of herding on ingestion and distance walked by sheep during spring (March-April), summer (June-July), and autumn (October-November) 1985" (El Aich and Rittenhouse 1988:279). Researchers divided sheep into three equal groups of approximately 40 ewes, 30 lambs and 4 rams. One flock, used as a control, was allowed to free graze. The other two flocks were assigned shepherds, one experienced, the other inexperienced. Each flock was given access to pasture approximately ten hours a day and then penned at night. Three sheep in each group were equipped with pedometers, while others carried vibracorders to record grazing time and periodicity.[7] Researchers took daily readings for each group.

---

7   Eating recurring at definite intervals of time

The results of the research showed that the novice shepherd's flock walked 11.4 km/day. In contrast the seasoned shepherd's flock walked only 6.3 km/day, a distance closer to that which the free grazing flock walked, i.e., 5.9 km/day. Each group spent the same time grazing, 508 +/- 42 min/day.

Grazing Behaviors

Grazing is comprised of several elements, including searching, selecting, harvesting and digesting food sources. Studies show that plant height and density affect the speed at which sheep graze. For example, sheep slow their grazing rate when plant height is above 2 inches. Moreover, sheep consume more clover than grass because they can eat it faster[8] (Lynch, Hinch and Adams 1992:21).

Sheep are generally more selective in their diet than cows because their muzzles are narrower. This allows them to choose which plant part they will eat first—usually the leaves[9] (Lynch, Hinch and Adams 1992:22-28). Sheep's previous experiences affect which plants they choose and they often learn how and what to eat from other sheep. For example, a friend bought a flock of yearling ewes from a large production sale in the fall. Unfortunately, these ewes had been raised on commercial feed. When spring arrived, my friend's flock was pastured on the same field as mine, but separated by fencing. My sheep continued to gain weight while my friend's flock lost weight. Because my friend's flock had never been out to pasture before, they had consequently not learned from other sheep how to graze. Not until they were pastured together with my sheep did they learn this skill and not until then did they begin to gain weight. Unfortunately, these sheep took a full year before they regained the weight they had lost.[10]

Heredity also influences grazing rate. In 1950, John Hancock studied monozygotic[11] cattle twins. He discovered the twins' grazing time was within 2% of each other, while the difference between the twin sets was 40% (Hancock 1950: 22-59). This means you should cull sheep that have poor grazing habits to increase your flock's overall grazing rate.

---

8 The average sheep will take 20,000 – 60,000 bites of food per day, depending on the type of feed (Lynch, Hinch and Adams 1992:22).
9 Leaf location varies with pasture maturity. Plants past their maturity will have dead, brown tops and green leaves at the bottom. Immature plants will have green leaves at the surface.
10 It is interesting to note that older ewes in a flock take even longer to adjust (Lobato, Pearce and Beilarz 1980:149-61).
11 Monozygotic refers to cows coming from the same egg, in other words identical twins.

Photograph 2.3 German Blackfaced sheep grazing (circular pattern) on heather in the Netherlands. At the time this photograph was taken, these sheep were tended by two dogs.

Grazing Patterns

Both scientific experiments and practical experience show closed flocks[12] rarely split into subgroups while grazing. Australian researchers combined sheep from two different breeds (Merino and Wilshire Horn) and studied the flock for nine days. During this time, each breed formed and maintained its own subgroup. After three months, they were confined to a small area, but, the researchers could still observe the two distinct groupings. The only time the two groups merged into a single flock was when the sheep were threatened by dogs. This has important implications for tending shepherds. Adding new sheep to a tending flock will cause two groupings to form within it. As long as these groupings remain, a shepherd and his dog must work harder to manage the flock (Winfield and Mullaney 1973:93-95).

Three different grazing patterns emerge when sheep free graze: circular (approximately 50% of the time), triangular and elliptical (Arnold and Maller 1985:183). See Figures 2.1 - 2.3. Too often, inexperienced shepherds allow their dogs to upset these grazing patterns by not holding their boundaries and/or racing around the grazing areas. When this occurs, sheep stand huddled in the middle of their grazing areas and are too afraid to eat.

Dogs are not the only factor that influences sheep grazing patterns. Sheep dispersal is influenced also by such factors as sheep breed, flock size and density, and food availability (Arnold and Maller 1985:173). Although sheep are flocking animals, the

---

12 Open flocks have members that come and go while closed flocks do not allow new members in. Many shepherds prefer closed flocks because they have endeavored to rid their sheep of a particular disease like sore mouth or OPP (Ovine Progressive Pneumonia).

Chapter Two: Sheep

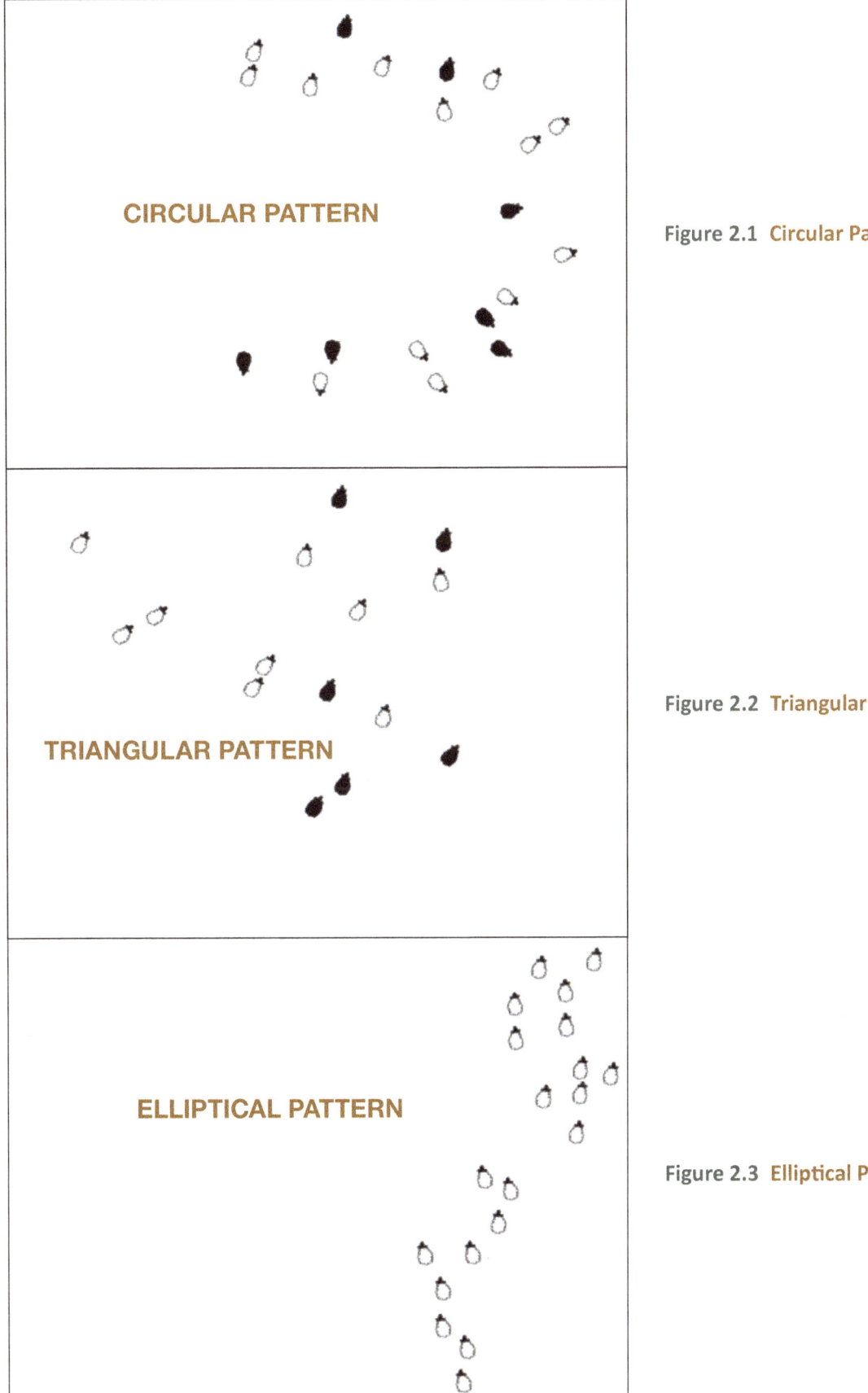

Figure 2.1 Circular Pattern

Figure 2.2 Triangular Pattern

Figure 2.3 Elliptical Pattern

degree to which they group together varies by breeds. For instance, the gregariousness of sheep ranges from high (Merino) to low (Scottish Blackface).

Individual and social distances play key roles in flocking rates. Sheep that have close individual and social distance are considered gregarious. Individual distance refers to the minimum distance animals approach one another. Social distance is the maximum distance a member moves away from another. Sheep breeds that have short individual and social distances are gregarious (Lynch, Hinch and Adams 1992:71).

Spatial differences occur in both gregarious and non gregarious sheep. Of the 311 main sheep breeds, 86 own Merino heritage (Mason 1969). These breeds are more gregarious than the other 225. However, spatial distances change throughout the course of the day for all breeds. For instance, sheep start closer together when they begin grazing after rest periods than after they have been grazing for a while (Arnold and Maller 1985:174).

Arnold and Maller measured spacial differences involving Merinos, Polwarths, Corriedales, Southdowns, Romneys, Border Leicester-Merino crosses and Horned Dorsets. The researchers pastured ten sheep of each breed in an area 5,926 meters square. They then took the following measurements of the area occupied by each breed of sheep: Merinos, 27 meters square, Polwarths, 27 meters square, Corriedales, 40 meters square, Southdowns, 33 meters square, Romneys, 54 meters square, Border Leicester-Merino crosses, 49 meters square, and Horned Dorsets 67 meters square (Arnold and Maller) 1985:176). Clearly different sheep breeds vary in area occupied.

Sheep also form pairs and these pairs lie at the center of the subgroups which form within flocks. The strongest bond is between a "ewe and her lamb, then between twins, filial[13] groups and finally, peers." The strength of the bond is measured by nearest neighbor distances (Lynch, Hinch and Adams 1992:69).

In another experiment, Arnold recorded the nearest neighbor distance of eight other breeds. They are as follows: Scottish Blackface, 7.5 meters; Welsh Mountain, 6.9 meters; North Country Cheviot, 5.5 meters; Dales Bred, 4.4 meters; Witshire Horn, 3.4 meters; Suffolk, 3.4 meters; Romney, 4.8 meters; and Merino, 3.1 meters (Lynch, Hinch and Adams 1992:73).

For some breeds, sheep numbers affect the grazing area occupied. Arnold and Maller found different breeds vary in spread (i.e., the distance individuals keep apart) and the number of sheep necessary to reach maximum spread. For example, it took ten Corriedales (non-gregarious) to reach the maximum spread for that breed, while it took 40 Merinos (gregarious) to reach theirs (Arnold and Maller 1985:187-188). Thus, when deciding how many sheep of a particular breed you need for training purposes, keep in mind that tending dogs were bred to cover great distances efficiently.

---

13   Previous offspring

Some tending enthusiasts in the United States believe they should train their dogs only with highly gregarious sheep because in Central Europe (the original home of the tending dogs) flocks are composed of Merinos or similar breeds of gregarious sheep. But, unlike Central European shepherds whose flocks generally range from 300-1,000 head, most shepherds in the United States keep flocks much smaller in size (on average, about 20). Unfortunately, shepherds who keep small flocks, should not train their dogs with a small flock of gregarious sheep. This is because gregarious sheep like Merinos will bunch together, taking up very little space. This causes dogs to move about very little. As a result, they will learn to work well at the head of the flock, but not at the tail.

By using this information, you can determine which sheep breed is best for you and the optimum flock size given your situation. For example, when I started training tending dogs, I used the sheep I had at the time (Cheviots). I found advantages and disadvantages to this breed. Their nearest neighbor distance is 5.5 meters, which meant I could use fewer sheep when training dogs for the wide and narrow grazes, but it took longer to train them and their wide flight zone was a disadvantage when working pens and roads.

A breed with a moderate level of gregariousness will probably work best for most tending shepherds. I found that Montadales worked particularly well. They have many of the same grazing characteristics of Cheviots, but they are more gregarious.

Topography and the availability of food also affect grazing patterns. For example, a study involving hill sheep in Scotland showed that during the winter, sheep group together more than in the summer when food is abundant. Moreover, sheep stay closer together after a rest period than after they have foraged (Arnold and Maller 1985:174).

Sample of Breeds in America (SIDS)		
Gregarious	Moderate	Non Gregarious
Merino	Columbia	Barbados
Navajo	Finnish Landrace	Cheviot
Panama	Montadale	Corriedale
Rambouillet	Polypay	Dorset
	Targhee	Hampshire
		Lincoln
		Oxford
		Shropshire
		St. Croix
		Suffolk

Sheep living in climates with hot summers and cold winters have distinct summer and winter grazing schedules. They generally graze for about eight hours a day (four hours after dawn and four hours at around sunset). During the summer, they graze roughly between the hours of 6:00 am and 6:00 pm. As temperatures grow colder, grazing schedules shift. Finally, in the depth of winter, the grazing period extends roughly from 4:00 am to 6:00 pm. (Lynch, Hinch and Adams 1992:13).

Maritime climates (e.g., New Zealand and Great Britain) have cold summers and an abundance of highly digestible grass and clover. In these climates, sheep generally eat nine times a day for 20-90 minutes and then rest for 45-90 minutes. Grazing generally stops at 10:00 pm and begins again at 7:00 am (Lynch, Hinch and Adams 1992:12-13). Sheep generally graze on the mountains in the summer and in the valleys in the winter. In New Zealand, fertilizing perennial grass in the valleys changed this pattern. This diminished the use of mountain pastures and, in turn, reduced erosion (Lynch, Hinch and Adams 1992:14).

Adaptability

Sheep forage on many different fodders, including grass, trees, shrubs, leaves and seaweed. On the Orkney Islands in Scotland, sheep have been confined to the coast line by a large stone wall since 1839. The sheep survive on a diet of course grass and brown seaweed. The tide controls grazing times and, as a consequence, most grazing occurs during low tide (Patterson and Coleman 1982:137-146; Lynch, Hinch and Adams 1992:11). Tides also control grazing schedules on the islands off the coast of Normandy. Grazing occurs there also during low tide. Then, as the water rises, the sheep gather together and a shepherd leads them to higher ground.

Food Preferences

Although researchers have conducted extensive work on the grazing behavior of sheep, very little was known about their taste and food preferences until the late 1970s. Hutson and Van Mourik (1981:575-582) offered 62 different foods to four sheep as supplements to their maintenance diet of hay. Their research discovered that sheep prefer grains and seeds over processed cereals, fruits, vegetables and sweets.[14] Among the top tested, sheep were attracted to barley, peas, wheat and a commercial horse ration. The most preferred grains were barley and wheat. Thus, both barley and wheat are suitable reinforcements to use in the development of a reward-based handling system for sheep (Hutson and van Mourik 1981:575).

Another study of food rewards and their effect on movement found that sheep trained on barley moved 90% more efficiently than unrewarded sheep. By spacing out training sessions to once every three to four days, it took less than a month to train

---

14   Processed cereals: Cornflakes®, Coco Pops™, Muesli, Wheaties™, Rolled Oats, Rice Bubbles™, All Bran®, Special K™, Sugar Frosties™, and Puffed Wheat™. Grain and Seeds: oats, barley, wheat, cracked corn, sunflower seed, rape, linseed, millet, tick beans, lupins, peas, crushed lupines, and sorghum. Sweets: sugar, raw sugar, licorice, chocolate, jelly babies, pop corn, and Smarties. Fruits and vegetables: sprouts, apple, celery, turnips, carrots, beans, potatoes, parsnips, silver beet, peas, cabbage, fennel, pumpkin, onions, cherries, and banana. Miscellaneous: Lucerne chaff, peanut meal, Blue Ribbon™ horse ration, Twisties™, potato crisps, milk powder, meat meal, starch, pea pellets, cattle cubes, oaten chaff, soybean meal, sheep cubes, bread and chaffed hay.

the sheep. In general, observers noted that it took sheep four to five days to become accustomed to all types of feed (Hutson 1985:264-273).

Water

An Australian study showed that sheep can exist for long periods without water, depending on the quality and abundance of forage. Researchers used two ranches from the same region as sites for the study. On the first ranch, salt bush comprised the majority of grass. As the name implies, such forage is high in salt content. On this ranch, sheep stayed within 7 km of drinking water and drank twice a day. On the second ranch, the sheep grazed on other plants. In contrast, this flock spent 2-3 days grazing in the summer heat before needing water. Only when the temperature reached 107° F for three consecutive days did the sheep need water daily (Lynch 1992:17).

We can conclude from this study that sheep receive water from the grass they eat. We also know animals eating large amounts of salt need more water. It is important to know and monitor the amount of salt your sheep ingest, whether it's in the grass they eat or in their commercial feed.

Even though sheep do not always need water everyday, they should be provided with daily access to it. Shepherds should make water available in the morning and at night. If dogs herd the sheep, however, they probably will not drink until later that night and perhaps not until the next day.

Movement and Rest

The environment plays a key role in sheep movement. In the mountains, cold air often settles in valleys and depressions at night, making higher elevations warmer. Consequently, mountain sheep move to higher elevations when it is cold to find warmer areas in which to rest (Geist 1971:272-275). This establishes a pattern. The sheep rest at higher elevations at night and in the morning move down to the valleys where they remain through mid-day. Then they begin their slow migration to the night-time resting site (Lynch, Hinch and Adams 1992:14). In warmer climates, however, sheep usually rest at night at the site where they stopped grazing (Lynch, Hinch and Adams 1992:18).

The speed at which different breeds of sheep move varies. Some breeds move about at an unhurried pace as they graze while others move swiftly from place to place. For example, "Norwegian sheep have a pattern of rapid intermittent grazing interspersed with trotting or running 10 to 50 meters before grazing again" (Lynch, Hinch and Adams 1992:14). Terrain also affects sheep movement. They move faster on flat than on sloping surfaces. When the texture of a surface changes sheep will stop and investigate before continuing forward (Lynch, Hinch and Adams 1992:87). When given a choice between moving up or downhill, sheep prefer to move up. They will move more easily into the wind than with it (The Sheepman's Production Handbook 1988: handling-6).

Resting

Before it lies down, a sheep will paw the ground using a front leg. Pawing removes debris i.e., rocks and twigs from the intended resting place. To lie down, a sheep lowers onto its front end, followed by its rear. During summer months, you can see sheep resting with their legs extended; in the winter, they tuck their legs closer to their bodies. (Geist 1971:259).

Erosion

Because sheep are creatures of habit, they can cause erosion. They like soft dirt to lie on and they will rest on the same spot over and over again. As they do, they paw the ground, which destroys vegetation. They also tend to take the same path to and from these resting sites. This creates winding trails, which make excellent paths for water to flow down after heavy rains and washes away needed top soil in the process (Geist 1971:275 and Ryder 1983:10).

Memory

Sheep are excellent at remembering both the good and bad treatment they experience for as long as a year (Hutson 1985). For instance, a predator invades your flock, chases your sheep and kills some of them. It may be a long time before your flock trusts a dog not to act like a predator. I recall a woman who entered a sheepdog trial and had the misfortune of exhibiting her dog on sheep recently attacked by coyotes. Unfortunately, her Belgian Shepherd acted and looked a lot like a coyote. Needless to say, the sheep responded with fear. It is interesting that these sheep distinguished between breeds. They reacted in terror only to the dogs that resembled coyotes.

Studies prove that lambs learn what types of food to eat from their mothers. Lambs removed from this type of feed for years still retain memory of it and readily eat it when reintroduced to it (Lynch, Hinch, and Adams 1992:28-32).

Vision

Sheep have a field of vision between 270° and 320°, depending on the amount of wool they carry (Sheep Production Handbook 1988: handling-3). Although they can see almost completely around themselves, their sight is most acute in the front (Kendrick 1990:62). Sheep's eyes are set on the sides of their heads, giving them both wide monocular vision (145° each eye) and narrow binocular vision (40° each eye).

Sheep are blind directly behind themselves for about 70° and in front of themselves about 2-3 cm (Lynch, Hinch and Adams 1992:36).

Although sheep can distinguish between people at short distances, at longer distances they "...cannot distinguish between humans, their sex, what they are wearing, whether the back or front view is presented or if the head and shoulders are covered" (Kendrick 1990: 64). Yet, they can see a person or a coyote approximately 1,000 yards away, even when partially hidden by shrubbery (Geist 1971:12).

Sheep have some color vision and can distinguish shapes[15] (Kendrick 1990:63, Geist 1971:12 and Seitz 1951:424-41). They also have astigmatic eyes and see vertical lines better than horizontal ones. This makes their vision less discerning than that of human beings (Kendrick 1990:63, Geist 1971:12 and Backhaus 1959:445-67).

Visual contact is very important to sheep. For instance, the winding pattern of sheep trails may be the result of the lead sheep's

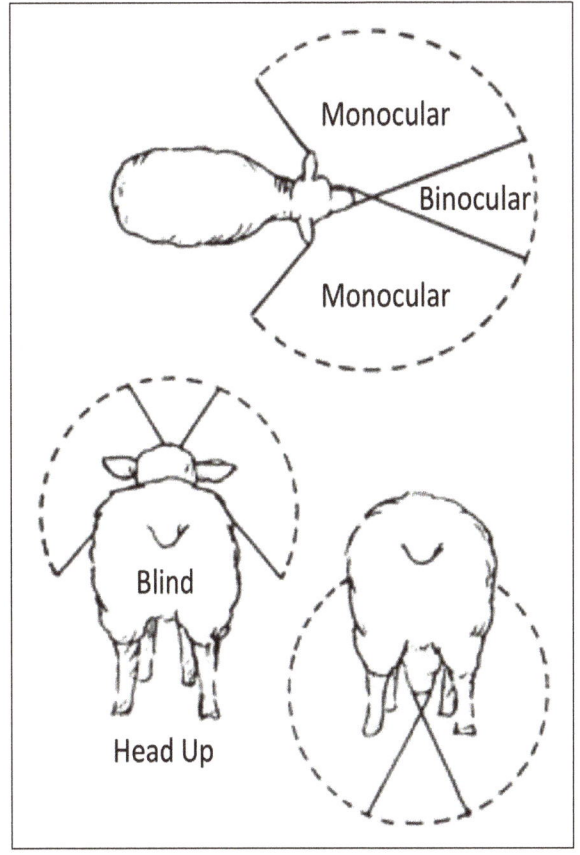

Figure 2.4 **Monocular and binocular vision (from Bill Kruesi handout)**

attempt to retain visual contact with the others following behind (Lynch, Hinch and Adams 1992:63). This is also why you must make sure that each sheep can see the one ahead of it when sorting sheep through handling equipment. This does not mean that chute sides should be see-through. On the contrary, chute sides should have a solid appearance. This will prevent the sheep from seeing each other and stopping as they travel through the system. Cutting gates should be constructed so the sheep can see through them. This will facilitate their movement into sorting pens (*Sheep Production Handbook* 1988: handling-6). Sheep do not like light contrasts. A lead sheep may balk at a shadow or a bright patch of sunlight and refuse to move into a pen or through a chute. This may be because their depth perception is better when standing still with their heads down than when moving with their heads up. "Sheep balk at shadows because they have to stop and put their heads down to determine the difference between a shadow or hole in the ground" (*Sheep Production Handbook* 1988: handling-4). This is why sheep jump over shadows. They think they are ditches or holes in the ground.

---

15 "Domestic sheep can distinguish a circle from a square, but not from a hexagon" (Seitz 1951); (Geist 1971:12).

Hearing

Because sheep are sensitive to loud noises, you should keep your voice quiet when commanding your dog (*The Sheepman's Production Handbook* 1988: handling-4). Your dog might obey your loud command, but your sheep may balk because they cannot tolerate your voice.

Vocal communication is very important to sheep. They use it when mating, identifying, warning or when distressed. "The repertoire of vocal sounds used by sheep ranges from a 'rumbling' sound made by ewes toward newborn lambs and also rams during courting, the 'snort' of aggression or warning to the 'bleating' of contact and distress call" (Lynch, Hinch and Adams 1992:52).

Each sheep breed makes distinct sounds. Researchers have discovered that sheep respond to the recorded sounds of their own breed faster than those of other breeds (Lynch, Hinch and Adams 1992:53). The most common use of vocal sounds occurs between a ewe and her lamb. Recognizing the voice of one's mother is very important in developing the bond between mother and offspring. The bond slowly declines with age, however. Because vocal communication is so important in the bonding process, it is important to listen to your flock and be able to mimic the voice of your sheep (See Chapter 3). One handler on the East Coast has perfected this ability. During competition, he calls to his sheep in a series of bleats. They respond much faster than his competitors' sheep. This is because his voice mimics the sounds of a mother ewe.

Aggression and Submission

Sheep also communicate with their bodies. This body language expresses a variety of moods. For example, aggressiveness is expressed by head clashing, butting, horn pulling, shoulder pushing and blocking. Head clashing ranges from two animals sparring by pushing each other with their horns to backing up and running at each other head on. Butting involves one sheep bashing its head into another's body. Generally, an animal is hit on the neck, side or rump. Horn pulling occurs when two sheep lock horns and pull sideways. Shoulder pushing is similar, but, instead of locking horns, the animals line up side by side and push each other, making contact at the shoulder. Blocking involves, nudging, horn and shoulder pushing and butting. This type of aggression is seen most frequently in sheep that are evenly matched (Lynch, Hinch and Adams 1992:56-62).

Threats take the form of kicks, horn threats, twists and low stretches, lifting heads and huddling. Sheep kick with their front legs. Generally, there is no contact with the intended opponent. In horn threatening, the sheep's head moves quickly downward and toward its opponent. Low stretches involve holding the neck out horizontally to the ground with the muzzle forward. The twist is added by turning the head 90° and flicking the tongue. Lifting the head is most often seen in rams pumping themselves up to look bigger by holding their heads as high as possible. Huddling is a threat response. The sheep form a circular group with their heads held low and close together.

Submissive sheep hang their heads low. They also back off as if in defeat. Smaller sheep will shake their heads in submission to larger ones. This is done after moving away from the larger animal (Lynch, Hinch and Adams 1992: 59-60).

Photograph 2.4 **Head clashing**

Photograph 2.5 **Butting**

Photograph 2.6 **Shoulder Pushing**

Photograph 2.7 Kicking

Photograph 2.8 Low stretching

Photograph 2.9 Submission

Flight and Fight Zones

Flight distance relates to the spatial behaviors described above. It is the distance a predator, i.e., a herding dog applies pressure before sheep will move away. Flight zone varies with breed of sheep. More gregarious sheep tolerate dogs working closer to them than less gregarious breeds. Because tending dogs work close to the flock, it is important to take flight distance into consideration when choosing the breed of sheep for your flock. Gregarious sheep require dogs that make close contact with them. In contrast, Border Collies were bred to work at a distance because they originally worked with Cheviots, a non-gregarious breed with a long flight distance. On the other hand, German Shepherds and Briards were bred to work with Merinos, a gregarious breed with a short flight distance.

Although the flight zone of sheep varies by breed, it is also affected by how they were raised. North Country Cheviots raised on the hills of Scotland have a bigger flight zone than Cheviots raised in close proximity to people and dogs. The closer the proximity, the smaller the zone. The distance ranges from 5 meters to 1 km (Lynch, Hinch and Adams 1992:82). Thus, sheep raised on the range in Montana will react faster than sheep raised in a petting zoo in New York City because frequent handling reduces flight distance. This can be used to your advantage when working with sheep that typically are not well suited for tending. With time and proximity, you can shrink the flight distance.

Photograph 2.10 **Sheep with a close flight zone**

The first phase in sheep's reaction to predators is attention-alert. During this phase, sheep freeze in place. They also stare alertly toward the perceived danger. Once this passes, the animals become alarmed, moving rigidly with their heads raised. Once a few individuals of a flock are frightened and flee from a perceived danger, the rest will follow quickly (Lynch, Hinch and Adams 1992:62). The degree to which sheep respond depends on both breed characteristics and conditioning to danger.

Following

One of the strongest instincts sheep have is to follow. Hundreds of years ago, shepherds capitalized on this trait to call their flock and lead them to pastures. This instinct is so strong that day-old lambs have been known to travel up to 2 km to follow their mothers (Lynch, Hinch and Adams 1992:171).

Sheep share this instinct with cattle. An article in the journal, "The Stockman Grass Farmer" told of a bull named "Ug" who was trained to come when called. When his owner called him, the rest of the herd along came. The farmer used Ug instead of people, dogs or horses to load his cattle. He even loaned Ug to other farmers to bring in their cattle as well ("Nation," 1991:1,8-11). When one sheep responds in alarm the rest soon follow.

Leadership

The tendency of sheep to follow a leader (as cattle followed Ug) is still utilized in many societies to control the direction in which the sheep move. A trained animal, often a wether or goat with a bell around its neck, is used as a leader. These animals are trained to follow a shepherd, accompanied by herding dogs, from one pasture to another (Lynch, Hinch and Adams 1992:85).

Many people confuse dominance with leadership. The leader is the sheep that initiates movement; the first sheep to move becomes the leader. However, the same sheep is not the leader all the time. Different sheep become leaders at different times depending on the circumstance. Differences in leadership also occur when sheep are forced to move rather than when their movement is voluntary. Quiet sheep tend to be leaders. In a test using groups of two-year-old ewes and three-year-old wethers, the quiet sheep of both genders led. Within the group of leaders, however, wethers led more often than ewes (Syme 1981:283-288).

At the end of the 1970s, labor costs increased in Australia's slaughter houses. As a result, more sheep needed to be handled by fewer people. At first, slaughter house personnel used dogs to herd the animals through their facilities. But the dogs caused damage by biting or jamming the stock into small areas so dogs were banned from the facilities (Bremner, Braggins, and Kilgour 1980:111). Quickly, the focus turned to the sheep themselves. The purpose was to develop sheep that:

1. Accepted a food reward when offered,
2. Were amenable to handling, remained quiet and calm in the presence of strangers with or without dogs present,
3. Could be led into and tied up in unfamiliar raceways, yards, wool sheds or trucks,
4. Would, on command, walk away from the handler toward a mob and then turn to lead it, and
5. Would walk at the head of the mob along a route which had either been learned or was indicated by open gates (Bremner, Braggins and Kilgour 1980:112).

The sheep used for this experiment were common hill sheep (three year old wethers) chosen for their leadership ability. This was determined by how they reacted both to food and human contact. They had to like barley and allow human contact in relatively small pens that reduced their flight zone (Bremner, Braggins and Kilgour 1980:112).

This experiment illustrates that sheep can be creatively controlled simply by taking advantage of their built-in follower and leader instincts. Thus, handlers do not have to rely solely on the predator-prey relationship between their dogs and the sheep to get the sheep moving. This method sometimes worries the sheep and exhausts the dogs. Instead, handlers often need only to observe the behaviors of their own sheep more closely to control the direction of the sheep.

Photograph 2.11 **A Dutch flock following its leader sheep. They are beginning to form an elliptical grazing pattern.**

Photograph 2.12 A Barbados sheep ram leading.

Photograph 2.13 A Tunis cross wether leading

Three

Shepherding

The Lord is my shepherd; I shall not want. He maketh me to lie down in green pastures: He leadeth me beside the still waters. He restoreth my soul: He leadeth me in paths of righteousness for His name's sake. Yea, though I walk through the valley of the shadow of death, I will fear no evil: for Thou art with me; Thy rod and Thy staff, they comfort me. Thou preparest a table before me in the presence of mine enemies: Thou anointest my head with oil; my cup runneth over. Surely goodness and mercy shall follow me all the days of my life: and I will dwell in the house of the Lord forever (Psalm 23, *King James Bible*).

To best use our dogs' herding capabilities, we must first understand the wider contexts in which those capabilities are rooted. We must understand, in other words, the agricultural practices in which raising, training and herding with dogs historically formed a part. This chapter begins with a brief historical sketch of shepherding before turning to shepherding practices themselves.

To understand shepherding, we must see it as a historical process. Many of its practices have changed through the centuries to meet new economic demands. Yet, the basic principles on which these practices rest remain unchanged. Some of the earliest written accounts of shepherding come from the Bible. Because early societies were so heavily dependent on pastoral pursuits, it is not surprising that shepherding themes abound in its stories, proverbs and psalms. We find many of the most informative of these in the stories and psalms surrounding the life of David.

Shepherding was always a solitary profession. Bound to their herds for weeks on end, shepherds in David's time traveled far from home to remote areas searching for choice grazing lands. In the process, they learned the most intimate details about their charges' behavior: they learned how to read their body language when a predator was near, where their sheep liked to lie down and why, how their sheep reacted to various forms of stress, and how to understand their communication. For example, a sheep's cry for salt sounds like a fluttered ru-ru-ru-ru-ru-ru-ru. By mimicking this cry, a shepherd could lead his flock quickly away from danger (Irigaray 1977:29). The water cry, in contrast, is a repetition of meee-maaa-meee-maaa (Irigaray 1977:27). "When the shepherd mocks this cry, he had better stand clear" for it can cause thirsty sheep to stampede (Baker 1987:17).

Shepherds carried a number of items of equipment: a rod, sling, staff, scrip, hog's oil, oil lamp and a small musical instrument such as a flute or harp. They used the rod and sling primarily as weapons of defense but they also used the sling to control the flock, much as shepherds in later times used dogs. If members of the flock roamed too far from the rest of the sheep, a shepherd would send a stone flying in the direction of the wandering sheep. It would not hit the offender. Instead it whizzed by its head, scaring it back toward the flock (Slemming 1942:50). Shepherds used their staffs to guide their sheep and pull down fruit that was out of their reach. Once harvested, shepherds fed the fodder to the sheep by holding it behind their backs as they walked ahead (Slemming 1942:46). Shepherds used a scrip (a bag made of goat skin) to carry their food and belongings.

A shepherd's day began with the rising sun, for if he started later in the morning, he would risk moving his flock in the heat of the day. At day break, a shepherd would step to the entrance of the sheep fold and call to his flock. Often, more than one flock spent the night together in a fold, but out of the rolling mass of animals would file all his sheep. None would remain behind. After this, another shepherd would come to the entrance and call his sheep using the same call. This would continue until all the sheep had left the fold (Slemming 1942:30).

As shepherds led their flocks away, they gauged their speed by the condition of their animals. If the flock had ewes with young lambs or pregnant ewes, it was important to move at a slow walk, for these animals could not keep a fast pace.

During the morning, shepherds allowed their flocks to graze on green, dew-laden grass. At noon-time, they led them to a cool location, such as the shelter of a rock. In the evening, they led them back to the fold to spend the night in safety.

Shepherds often found the best grazing on mountains. As they led their flocks to the mountain pastures, they knew their sheep would need energy to make the climb. To prepare for the journey, they often stopped their flocks in lowland pastures to graze and rest. They also stopped beside running water. Knowing that sheep cannot drink from fast running streams, shepherds used their staffs to divert water into calm pools for their flocks to drink.

Many dangers threatened the flock. For example, as they approached grazing areas, shepherds carefully searched the tall grass with their staffs, looking for adders living in holes. The snakes could suddenly slide out and strike the sheep on their noses. Consequently, shepherds first placed their flocks at a safe distance and then studied the terrain carefully. When they located all the adder holes, they would pour a circle of hog's oil around each hole. The adders' smooth bodies lacked the friction to slide past the oil to strike the sheep. Since sheep find hog's oil repulsive, they avoided the snake nests as they grazed (Slemming 1942:46-47).

During biblical times, many areas in the Near East were overgrown with thorns, briars and thistles. In order to pass through, shepherds walked ahead of their flocks using their rods to beat paths through the brush. Nevertheless, lambs sometimes became entangled in the bushes. When shepherds spotted these lambs, they gently lifted them from the snares with their staffs.

In the evening, shepherds returned their flocks to the sheep folds. Sheep folds were constructed simply, consisting of four low walls with an opening at one corner. A covered area inside the fold was used during severe weather. As a shepherd approached a fold, he would stand at the opening and tap the back of each sheep as it passed by, watching for hurt or sick members of his flock. He would also take a head count, assuring himself that all his sheep were present. If one was missing, he would ask another shepherd to watch over his flock while he searched for it. If all was well, the shepherd would lay down for the night at the opening of the pen that contained his flock. From there, he could react quickly to any danger that threatened his sheep.

This chapter began with the 23rd Psalm, the psalm of David. In composing it David, explained the kind of relationship he believed existed between man and the supernatural. To make his explanation clear, he developed a cultural metaphor based on his earlier life as a shepherd, the kind of life outlined in this chapter. In this metaphor, God stands as the shepherd and mankind as the sheep. The relationship is described from the perspective of the "sheep," i.e., the human voice is in the first person. This metaphor made sense to David and his contemporaries because the pastoral way of life was so deeply ingrained in their lives. They had a solid understanding of shepherding because it was critical to their survival. Thus, David could build from common knowledge to help others share his understanding of the world.

Pastoralism, however, is not as critical to the survival of our society as it was to David's. Consequently, most of us did not grow up as shepherds, so we must learn good shepherding the same way we learn a second language as an adult. The 23rd Psalm gives us an excellent framework in which to accomplish this (regardless of what its ultimate truth may or may not be) because it clearly articulates the fundamental and unchanging principles underlying effective shepherding practices. In the pages that follow, we break the psalm into the principles directly related to shepherding and then expand on these principles. Here, however, we switch perspectives in that we are the shepherd, not the sheep.

"The Lord is my Shepherd; I shall not want"

Good shepherding is being sensitive to the needs of your flock. Not only must handlers be knowledgeable dog trainers, they must also have an intimate knowledge of and feeling for their flock. They must spend as much time observing their flock and building a relationship with their sheep as they spend building a bond with their dog. Prospective shepherds may rent a flock by paying a lesson or entry fee at a herding event; they may also buy a flock of their own from a sheep producer. Regardless, they own the sheep for a specific amount of time and it is their responsibility to care for the sheep while they "own" them. Although we pay entry fees at herding trials and do not get to pick the sheep we want, we still must treat these sheep as our own. An entry fee gives us more than just a chance at running our dogs. It is the price we pay for the responsibility of becoming shepherds.

When people who don't own and care for sheep themselves become involved in herding, often their sole desire is to train their dogs and enjoy their sport. Such people sometimes see sheep as a tool to be used toward this end. For example, at the end of a herding event, a man approached me and asked how much money it would take to allow his dog to work my sheep. Money was no object, he said.

This man saw the sheep as a means of working his dog and he was correct. Nevertheless, his perspective was wrong. The sheep were tired from a long day of work and needed to rest. I told him I would not allow him to use my sheep for any price but I would sell them to him. In that case, however, I would not take them back. I was not willing to let someone who didn't care for them as his own use my sheep, no matter how much money he offered. He declined to pay because he did not want the responsibility that ownership entailed. Ownership alone does not make people good shepherds. It just makes them owners. Good shepherds love and care for their flocks. They spend time, money, strength and sometimes even give their lives to provide for the needs of their sheep.

The needs of sheep are quite simple: food, water, shelter, and protection. Philip Keller, the author of *A Shepherd Looks at Psalm 23* and *A Shepherd Looks at the Good Shepherd and His Sheep* explains this in a particularly moving way:

"The tenant sheepman on the farm next to my first ranch was the most indifferent manager I had ever met. He was not concerned about the condition of his sheep. His land was neglected. He gave little or no time to his flock, letting them pretty well forage for themselves as best they could, both summer and winter. They fell prey to dogs, cougars and rustlers.

Every year these poor creatures were forced to gnaw away at bare brown fields and impoverished pastures. Every winter there was a shortage of nourishing hay and wholesome grain to feed the hungry ewes. Shelter to safeguard and protect the suffering sheep from storms and blizzards was scanty and inadequate. They had only polluted, muddy water to drink. There had been a lack of salt and other trace minerals needed to offset their sickly pastures. In their thin, weak and diseased condition these poor sheep were a pathetic sight.

In my mind's eye I can still see them standing at the fence, huddled sadly in little knots, staring wistfully through the wires at the rich pastures on the other side.

To all their distress, the heartless, selfish owner seemed utterly callous and indifferent. He simply did not care. What if his sheep did want green grass; fresh water; shade; safety or shelter from storms? What if they did want relief from wounds, bruises, disease and parasites?

He ignored their needs—he couldn't care less. Why should he—they were just sheep—fit only for the slaughterhouse" (Keller 1970:28-29).

In summary, a good shepherd lives, works, and in some cases dies to see his sheep prosper because their contentment brings him great joy and pride. His flock could give him no greater gift in return for his labor than to prosper.

"He maketh me to lie down in green pastures"

So, good shepherds provide for the needs of their flocks, but this is not always as simple as it sounds. Take forage, for example. Some shepherds can provide their sheep with good grazing by taking them to nearby mountain tops or open woodland meadows where green grass naturally grows. Others, however, are not so fortunate and must create their own pastures. These shepherds must clear their land of brush, trees, roots, stumps and rocks. And once the land is cleared, the soil must be plowed and tilled to make a soft bed in which to plant the grass seed. Their work is still not finished, however. They must make sure the newly planted seedlings have adequate water. If the rainfall is inadequate, the shepherds must install irrigation systems to water their grass crop. All this labor must be undertaken so that their sheep's needs are met.

Needs are not always identical to wants. Sometimes, sheep want what they don't need and want what will do them harm. Good shepherding requires you to assess situations carefully and make judgment in terms of the well being of your flock, regardless of what your sheep desire.

In this respect, sheep are like people; they often desire more than they need. Even when their needs are met, they sometimes crave more and set out to find it. For example, seeing green grass growing on the other side of a fence may cause the more adventuresome sheep to become discontent and follow a fence line for a place to break through. Unfortunately, the rest of the flock usually will follow. If successful, the flock escapes the confines of its pasture. Unfortunately, it exposes itself to danger in the process.

Even if the flock cannot escape, the discontent of the adventuresome sheep can be contagious, making it impossible for the other sheep to graze peacefully. To combat this problem, shepherds hang bells around the necks of the adventuresome sheep, so that as the sheep walk, the bells ring. Attentive shepherds can distinguish the different rhythms of the bells; the more frequently the bells ring, the more restless the sheep. When they hear "restless" ringing, good shepherds take steps to calm the flock.

Unfortunately, sheep prone to discontentedness can pass this characteristic to their off-spring. For this reason, they should be sold. Otherwise, you can breed this characteristic into your flock with serious, negative consequences. For example, I owned a ewe that escaped our pastures continually and took the rest of the flock with her. Unwisely, I chose not to sell this particular ewe because her conformation was exceptional and I thought she would produce nice lambs.

And she did. Unfortunately, their temperament was just like hers; they were always escaping. Finally, I had to sell this ewe and her offspring. If I had not, I soon would have had too many Houdini sheep, each one an escape artist. The entire flock would have suffered as a result.

As another example, sheep enjoy eating and then lying down. Like people, they often will not lie down voluntarily and rest when they need it. Good shepherds know the limitations of their flock and know that when their sheep over-extend themselves, they risk injury and even death. Good shepherds often force their sheep to rest for their own good.

Besides hunger and thirst, sheep won't rest for two reasons: fear of external threats (which we'll discuss later) and competition for dominance within the flock. After a flock settles into an area to graze, competition for the best grazing begins. Head butting starts as new members strive to establish their position in the hierarchy of the flock. Constant jockeying for position creates turmoil. As a result, no one in the flock can get peace and quiet. Therefore, good shepherds take necessary steps to establish calm in the herd. Sometimes, just walking among the flock will stop the fighting. Other times, shepherds must discipline unruly members by tossing clumps of dirt or small stones at them or using harsh tones of voice.

"He leadeth me beside the still waters"

A sheep's body (as ours) is composed of roughly 70% water. Without enough water, sheep become dehydrated, weak and finally die. When sheep become thirsty, they begin to search for water. If clean water is not available, they will drink dirty water. Unfortunately, parasites and potentially harmful bacteria and viruses live in polluted water. While attending a herding competition in Germany, I saw good shepherding in practice. A large puddle of dirty water was located on the roadway and each time shepherds moved their flock past it, they placed their dogs at the puddle to keep the flock from drinking. Even in competition, they tended the sheep as if they were their own.

Good shepherds provide their flocks with a regular supply of water. Sheep get water from a number of sources: wells, dew, streams and from the moisture in grass. When we pasture sheep during the summer, we must bring water to them because the only natural water supply comes as early morning dew. If the weather is cool, our sheep do not drink the well water we provide; they can live for several months on the moisture in grass without drinking from their water troughs. Many people are surprised that the amount of water we transport is small compared to what our sheep need in the winter when they feed on baled hay. This is because hay has very little moisture content.

Salt also increases sheep's need for water. During lambing season, we separate pregnant ewes and ewes with lambs from the rest of our flock by putting them on their own side of the barn. One night, we found all our flock on the same side. The sheep that changed sides had to climb and trample a section of cattle panels, a task not easy to accomplish. At first, I had no idea why they tore down the fence. Later, I realized one group of sheep had consumed more salt than usual. Once the sheep emptied their trough, their thirst drove them to break down the barrier in search of more water.

Finally, flocks pastured in the mountains must get their water from streams. In the mountains, water runs fast, especially in the spring or after a storm. Although this water is fresh and clean, sheep will not drink from it because they prefer clear, still water. Sometimes, they can find a quiet pool of water. Usually, however, they must rely on their shepherds to make one by digging a small trench that empties into a deeper pool (Slemming 1942:43).

"He restoreth my soul"

Even a good shepherd's flock can develop health problems. When sheep are not well, they may lie down, refuse to eat, or separate themselves from the rest of the flock. Their feces may look different, their breathing may be faster or slower than normal, their wool may be brittle and they may have fevers.

Although most health-related tasks involving sheep are preventive, e.g., vaccinating and worming, shepherds still need a basic knowledge of first aid. They may need to suture a laceration, inject an antibiotic or deliver a lamb. At first, these procedures may seem difficult, but with practice, they become second nature.

Your local feed store or supply catalog offers everything you need to care for your flock. If a member of your flock has a cut or infection, give it an injection of antibiotic. If it's just a small cut or abrasion, apply a topical antibiotic ointment or spray to reduce the chance of infection. If a sheep has a laceration, it may need to be sutured. Several books are available dealing with such health problems: *Raising Sheep the Modern Way* by Paula Simmons and *The Sheepman's Production Handbook* published by the Sheep Industry Development Program are two helpful guides. No shepherd's library is complete without them.

Not all methods of restoring health involve first aid. For example, a few years ago, I visited Geoff and Viv Billingham in Scotland. The Billingham's are well-known shepherds and competitors who pasture their sheep in the Cheviot Hills. One day, as I walked with my dog on the hills far from their cottage, I noticed a ewe cast (lying on its back with its feet pointed upward) at least a half mile away. Because a steady rain was falling and I wanted to get back to the cottage for tea as soon as possible, I sent my dog out to her. I hoped this particular sheep would see my dog and get up. Unfortunately, she refused to move and I had to go to her to set her on her feet. When I told Viv what happened, she said that school buses carrying school children stop if anyone sees a sheep "cast" on the hills. Even school children learn to care deeply for sheep. As it turned out, this was one of my most rewarding experiences working with sheep. Since then, I have had many opportunities to help my own Cheviots back on their feet.

Cheviots are not the only sheep that become "cast." Any sheep can find itself in this same position. Being overweight, carrying a heavy fleece or just finding a comfortable spot on the ground may cause sheep to become cast. Unfortunately when sheep are on their backs, gases in the rumen increase to the point that they become like a tourniquet, and cut off circulation in the legs. This makes it almost impossible for sheep to stand. Their legs become numb and they can die.

"He leadeth me in paths..."

Flocks in the East are generally led or guided from pasture to pasture, while flocks in the West are not (Baker 1987:45). When a shepherd guides his flock from the front, he can see hazards and find good fodder more easily. This enables him to change course when needed. When a shepherd walks behind his flock, he cannot anticipate hazards as easily. While in the Netherlands, I asked a shepherd whether he walked in front or behind his flock as they moved from pasture to pasture. He answered that in the morning he walked in front of the flock, so that he could choose the best pasture, but at night, he and the sheep were tired. He walked behind them because they knew their way home.

I asked another shepherd the same question. He said he always walked in front of his sheep, even when tired. Soon afterward, we saw a snake blocking our path. The shepherd quickly stopped his flock and placed his dog as a pick,[16] so the sheep would avoid the snake. If this shepherd had not been walking in front, he would not have seen the snake and his sheep and dog might have been bitten.

Sheep are creatures of habit; when they find an area they like, they wear it out. If you walk into a pasture, you will find areas the sheep have worn bare. Because such areas are conducive to parasite infestation and erosion, both your animals and the land suffer. The best defense is to keep your flock on the move, which keeps overgrazing and parasite infestation to a minimum. Planned rotation of flocks from pasture to pasture has similar results. Karl Fuller, a professional shepherd and well-known trial competitor in Germany, leads his flock of 800 sheep to a different grazing location each day. Some days, he walks three miles, other days, six or even more, depending on where he can find the best grazing. I asked Karl about parasite problems, but because his flock moves so often, he seldom has one.

"Thy rod and Thy staff…"

Throughout the world, shepherds use specialized equipment in their work. Sometimes, they carry backpacks or bags to hold necessary items, i.e., food, water and medical supplies. Today, some shepherds carry rifles. Traditionally, shepherds carried rods and staffs (crooks). Even today, crooks are an essential piece of equipment for most shepherds. Two basic types of crooks exist: neck and leg crooks. Neck crooks are usually curved at the top end. The diameter of the curve depends on the size of sheep tended. A shepherd raising Cheviots in Scotland will have a crook with a smaller diameter to its curved end than British shepherds raising Suffolk sheep.

To use a neck crook, shepherds hold the crook by its straight end and as the sheep they wish to catch passes by, they hook it by its neck. These crooks are usually made of ash, chestnut or hazel. They often have attached sheep's horns that serve as handles for walking in the hills. In Germany, the crook is called a shepherd's spade. It, too, is curved at one end, but unlike the shepherd's crook of Great Britain, it has in addition a shovel and leg hook fastened to the straight end of the crook's shaft. Shepherds use the shovel to pick up dirt and toss it at the sheep if they get out of line or move too quickly. Shepherds use the leg hook to catch sheep by their hind legs.

---

16 A pick is when a dog establishes a stationary position for a specific purpose. In the example of the snake, the dog is used to block the sheep from moving into a danger zone.

Photograph 3.1 Dutch shepherd's staff. It does not have a leg or neck crook. The shovel sides are bent to form a cup because the ground is very sandy.

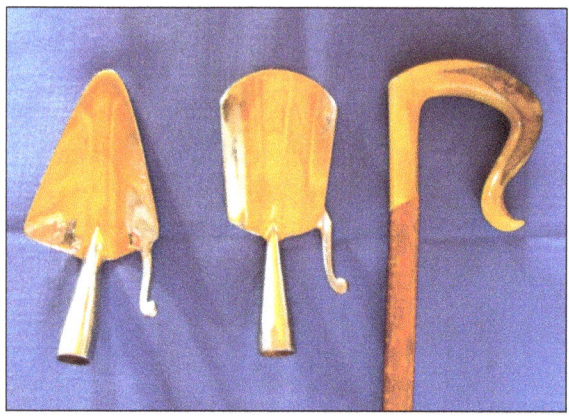

Photograph 3.2 Two German shovels, one for hard ground and the other for soft (similar to the Dutch shovel) and a Scottish shepherd's staff. Notice the staff's sheep horn handle and size.

Shepherds throughout the world often say their crooks are extensions of themselves and would not consider working a flock without these "right arms." Unfortunately, people in the United States have no idea how to work with these tools and use them only for training purposes.

"Thou preparest a table before me"

Preparing a table for a feast requires work and planning, but preparing a "table" for a flock of sheep takes even more work. The average sheep needs four pounds of hay per day (1,460 pounds per year). This means, each sheep needs 36.5 bales per year because the average bale of hay weighs 40 lbs. Thus, a flock of 70 needs 2,555 bales of hay per year.

Preparing a hay field takes time and labor. You must kill the weeds, take soil samples, apply lime, plow the soil, choose grass seed, sow the seed and apply fertilizer. That is just to grow the grass! The next step is to cut, rake, and bale the hay, which then must be stored in a dry place. Fortunately, many shepherds can pasture their flocks in the spring, summer and fall, so their sheep require less baled hay to eat.

Pasture areas must be prepared also. Fence lines must be checked for damaged sections and repaired. Water sources must be inspected and any debris cleaned from streams or water troughs. If shepherds are using new grazing areas, they must first remove all brush and poisonous plants. (See Appendix II for a list of poisonous plants).

Grass Management

Because of rising feed costs, many sheep producers farm grass, as well as shepherding their flocks. Without grass farming, feed consumes fifty to seventy percent of a flock's total feed costs. When sheep harvest their own forage, shepherds decrease feed costs and increase profits (Sheepman's Production Handbook 1986: Man 31).

The key to successful grazing management is "… to increase animal production while maintaining or improving the quality of the pasture" (Kruesi 1985:29). Good grazing management, therefore, requires understanding the relationship among plant growth, sheep distribution and stocking rates. Because overgrazing weakens plants and slows their growth, shepherds must determine carefully the optimal number of animals to graze in any given area and the best grazing schedule given, of course, the nature of the plant cover.

Admittedly, this is not an easy task because plant growth changes from season to season and year to year. Generally, most plants grow more quickly during the spring and summer. To avoid overgrazing, harvest excess plant material during rapid growth periods and feed the excess fodder to your sheep during slow growth periods. According to *The Sheep Production Handbook*, you should forage only about half the total estimated growth for a given growing season. Still, during the dormant stage of growth, you can harvest 65% of the plant cover without causing serious damage. Of course in times of drought you must reduce the size of your flock (*The Sheepman's Production Handbook* 1986: Man 31).

Sheep distribution refers to how uniformly your flock forages different sections of a grazing area. Generally, sheep will spend more time grazing near their water supply. The type of vegetation and the steepness of the land also cause different breeds to concentrate grazing in particular areas (*The Sheepman's Production Handbook* 1986: Man 32). For example, when hill sheep (Cheviots) and lowland sheep (Dorsets) are grazed together on a mountain side, the Cheviots tend to graze along the steeper slopes, while Dorsets usually graze near the base of the mountain.

Stocking rates range from light to heavy. When rates are light, the number of animals is too low to consume the available forage. In such situations, sheep selectively eat the more palatable plants first. This allows the sod to open up, which, in turn, causes the most desirable grasses to cluster, leaving open areas for weeds to grow. Animals gain the most weight when stocking rates are light because they can graze selectively the best forage in the pasture. Yet, light stocking rates under utilize a pasture's resources because the stock leave some forage uneaten.

When a pasture has a high stocking rate, the animals are not as selective in what they eat, since each animal must compete with the next for available forage. As a result, plants grow at a uniform rate, allowing legumes to compete for sunlight on an equal basis with other grasses. A stocking rate that is too high, however, causes overgrazing. This can defoliate plants, so that they must draw on their root reserves to survive (Kruesi 1985:33-34). When plants must use their reserves, they do not flourish. This causes plant growth to slow and the pasture's available nourishment value to diminish. Therefore, a wise shepherd will remove his flock from a grazing area before the plants must draw on their root reserves.

Grazing Systems

A well planned grazing system permits a pasture to improve, thus increasing livestock production (*The Sheepmans Production Handbook* 1986: Man 34). Because no grazing system works proficiently under all conditions, shepherds must chose the one that best fits their needs. Agricultural experts have developed several types of grazing systems, including continuous, short duration rotational, alternate and strip grazing.

Continuous Grazing

Many sheep producers in the United States use continuous grazing to feed their flocks. Once the spring growth is established, shepherds take their flocks to pasture where they stay until autumn and the grass stops growing. Unfortunately, when grazing is continuous, sheep overgraze the plants they like best. They ignore the less palatable ones or trample them down. When the palatable plants are overgrazed, they die out. As a consequence, less desirable plants generally replace them and the quality of forage degenerates.

Short Duration, Rotational Grazing

In short duration, rotational grazing (also called intense rotational grazing) the flock is concentrated in several small grazing areas or cells, thus increasing the stocking rate. This forces the sheep to graze the plants evenly, which, in turn, reduces weed growth. Each cell is systematically grazed. "Livestock are allowed to forage in each area just long enough to harvest the existing forage, but not long enough to eat any regrowth" (USDA Soil Conservation Service: nd). For instance, orchard grass should be grazed when it reaches a height of 10 inches; grazing should cease when it is reduced to 3-4 inches. Blue grass, in contrast, should not be grazed until it reaches a height of 6 inches and then grazed until it is reduced to 2 inches (USDA Soil Conservation Service: nd). Once the flock harvests the available forage, it moves to the next cell and does not return until new forage has grown. Compared to continuous grazing, the results are:
1. A grazing period extended four to nine weeks;
2. Increased dry matter yield; and 3) high quality feed (University of Vermont Extension Service: 1984).

The size of each cell should vary depending on the rate of seasonal regrowth and stocking. During the spring, when the grass grows quickly, you can place more sheep in a given area than in the fall. Therefore, the best ratio of sheep to grazing area size fluctuates throughout the grazing season. As a general rule, ten sheep weighing approximately 100 lbs. each need one acre of pasture (USDA Soil Conservation Service: nd).

Alternate Grazing

Alternate grazing lies between continuous and short duration, rotational grazing. The grazing area is divided into two cells and grazing alternates between them. Animals generally spend 2-3 weeks in each cell. Because there are only two grazing areas, it is impossible to keep the sheep from regrazing the preferred grasses before they have recovered fully. As a result, the pasture is unevenly grazed, leaving an unmanicured appearance. More importantly, the nutritional value of the pasture decreases. In addition, sheep move about the pasture searching for choice forage instead of conserving their energy, which reduces weight gain and profits (Kruesi 1985: 32-33). Although alternate grazing better utilizes the pasture than continuous grazing, it does not work as well as short duration, rotational grazing.

Strip Grazing

Strip grazing is a very intense form of short duration, rotational grazing. The shepherd keeps his flock in a cell for only one day before moving the sheep to the next cell. This system works very well in pastures with tall, high-yielding forage, i.e., sudan grass and millet or annual crops, such as corn and turnip. When row crops are foraged, the shepherd allows the flock to graze only a few rows at a time. This keeps the sheep from trampling unharvested forage. Generally, this type of grazing is done in pastures where flora easily flourish.

Short duration, rotational grazing in any form, has a number of useful consequences: Suppose you divided an acre of pasture into ten sections and grazed 20 sheep on each section for three to four days. This would return the sheep to each area grazed every 36 days and, as a result, minimize worm infestation, reduce the amount of necessary pasture and extend the grazing season. When we first instituted intense rotational grazing, we extended our grazing season and cut enough hay to feed our sheep until they went back to pasture the following spring.

Most farmers use permanent or temporary electric fencing to create cells. By using dogs trained in the tending style, shepherds do not need the electric fencing. They do, however, need to supervise their dogs.

The following books dealing with grass management techniques are available from the journal, *The Stockman/Grass Farmer,* Mississippi Valley Publishing Corp. Jackson, MS., (601) 853-1861 or 1-800-748-9808:
- *Grass Productivity,* André Voisin
- *Greener Pasture's on Your Side of the Fence,* Bill Murphy
- *Manual of the Grasses of the United States,* A.S. Hitchcock
- *How to Plan Implement and Practice Controlled Grazing on Your Place,* Bob Kingsbery
- *Understanding Grass Growth,* George Gates

"mine enemies"

Shepherds must be unerring in protecting their flocks because sheep have many enemies: predators, poisonous plants and parasites, to name a few. The most widely recognized killer of sheep, however, is the wolf and in North America, the coyote. For years, North American farmers and ranchers killed wolves and coyotes with poison, exploding bait, and fire arms. They also burned tires, set traps, played loud music, and discharged explosives to drive the predators away.

Unfortunately, many of these methods are environmentally unsound. For example, ranchers widely used Compound 1080 to control predators. They laced carcasses with the poison and left them for the coyotes to eat. In 1972, Compound 1080 killed approximately 100,000 coyotes (National Wildlife 1986:16). Unfortunately, bears, eagles and other wildlife also ate the poison and died. Even though Compound 1080 killed many coyotes, it did not stop coyotes from killing millions of sheep throughout North America. Between 1972 and 1973, 55% of the lambs that died in New Mexico were killed by coyotes (Anderson, Hulet, Shupe, Smith and Murray 1988:252). Because Compound 1080 was neither environmentally safe nor effective in minimizing sheep losses, the government banned its use in 1972 (National Wildlife 1986:16).

After Compound 1080 was banned, farmers and ranchers began to explore more environmentally safe techniques of predator control. In 1985, R.H. Blackford noted that sheep losses decreased when sheep and cattle grazed together (Blackford 1985:204-206) For this to be effective, the sheep and cattle must bond in order for them to graze close together (Anderson, Hulet, Shupe, Smith, and Murray 1988: 251-257). Shepherds have tried other methods also with varying degrees of success, e.g., electric fencing flock guardians: alpacas, llamas, donkeys and guard dogs.

Using guard dogs as flock guardians has a long history in Eurasia. Their use is reported in the Bible's Old Testament (Job 1:1-3, 30:1) and in *De Ra Rustica,* written circa 60 A.D. In North America, Navaho shepherds have used mixed breed dogs successfully to protect their sheep for years (Downs: 1964). In the 1970's, Ray and Lorna Coppinger began a twenty-year study regarding the effectiveness of European and Asian guard dogs for predator control. As a result of this research, American farmers and ranchers began using guard dogs to protect their flocks. The dog's natural instinct to bond with their flock (to think of the flock itself as their pack) makes them very effective against predators. Their use is an option every sheep producer should consider.

Poisonous Plants

Some plants are deadly to sheep. When sheep die from an unknown cause, a poisonous plant is often the culprit. Environmental conditions can play a key role in causing plants that normally are nontoxic to become poisonous (while sometimes causing poisonous plants to become less toxic). For example, freezing, wilting or crushing a normally nontoxic plant can cause it to become toxic (*The Sheepman's Production Handbook* 1988: Health 54-55). Young plants growing in the shade can be more poisonous than adult plants growing in the sun. Finally, the amount of water an animal consumes can increase the speed at which cyanide poisoning works, which is why many sheep poisoned by hydrocyanic acid are found dead near water sources (*The Sheepman's Production Handbook*: green cover 3-72).

Sheep that are range grazed should be fed a salt and phosphorus supplement for two weeks prior to being pastured (*The Sheepman's Production Handbook*). This is because arid ranges are drier than most green pastures. Thirsty sheep are often salt hungry, which leads them to consume some plants they would normally avoid, e.g., selenium, various astragalus species, halogeton and other similar toxic plants. Moreover, sheep deficient in phosphorus often eat plants and even dirt they usually avoid. This also increases the chance they might consume poisonous plants (Woody Lane: personal communication).

You can protect your flock from poisonous plants by following a simple set of rules:
1. Identify the poisonous plants in your area. Usually, only 4-10 species of poisonous plants cause major losses in each grazing area (*The Sheepman's Production Handbook*: green cover 3-74.
2. Learn to identify these plants at each stage of growth.
3. Inspect your grazing areas for poisonous plants and remove them, if possible, before allowing your sheep to graze.
4. Don't allow thirsty or hungry sheep to graze in areas where toxic plants grow.
5. Supplement your sheep's diet with salt and phosphorus.
6. If you must graze your sheep in areas infested with poisonous plants, do so when the plants are less toxic.
7. Make sure sheep have access to water at all times, so they will not search out other sources, i.e., poisonous plants they would otherwise not eat.
8. Learn to recognize signs of poisoning (*The Sheepman's Production Handbook*: green cover 3-74) See Appendix II for a list of poisonous plants.

Parasites

Parasites are of two general types: internal and external. According to Kruesi, sheep are more susceptible to internal parasites than cattle because they can graze closer to the ground and their manure is pelleted. Thus, sheep can graze over their fresh feces while cattle cannot. This is because cattle manure blocks access to the grass beneath it (Kruesi 1985:69).

As sheep graze, they become infected with parasitic worms by eating their larvae. When the larvae become adults, the females produce eggs, which are passed by the host in its feces. Once outside the host's body, the eggs grow into larvae and the process repeats itself.

The life cycle of worms differs, but those that cause problems in sheep are similar, about 21 days. In winter, worms go into dormancy. During this time, ninety-five percent remain in your sheep, while five percent lie dormant in your pasture. In the spring, the numbers reverse (Simmons 1989:162). Because some medicines are ineffective against dormant worms, it is important to choose medicine that is effective against such worms when worming sheep in winter.

Parasites and anthelmintic resistance are serious problems in sheep and goats. I recommend that you consult your veterinarian regarding a customized plan for parasite control. Keeping sheep off a grazing area for a year breaks the worm cycle and makes the pasture relatively worm free. This does not mean, however, that the pasture must lay fallow. You can grow other crops or graze cattle on these pastures because parasites are specific to their host. You can alternate sheep and cattle on a field from year to year.

When lambs are pastured with their mothers, they generally remain worm free for about a month. Therefore, worm your sheep again at this time to kill the parasites picked up during the first month of grazing. Finally, worm your sheep the day you take them off pasture for the winter (Kruesi 1985:70).

Even with careful attention to proper worming protocol, a flock can still become worm infested. For example, in the past, I wormed four times a year. A few years ago, I had four lambs die very quickly. Extensive laboratory testing determined the lambs died of coccidia infestation even though they had been fed medicated grain. So I added Co-rid, a stronger coccidia medication to their water. I still lost three more lambs. I then added a third oral medication to their diet, which finally broke the death hold on my flock. Why did a problem arise that particular year? Sheep thrive best in semi-arid conditions. That year, we had a rainy spring and the pasture was very wet. This made the grass grow so quickly that the sunlight could not kill the parasites living in the ground. Consequently, my sheep needed more medication. Now, during high grass growth periods and when my flock is contained for relatively long periods of time, I worm on a tighter schedule.

In the past, shepherds moved their sheep from pasture to pasture seasonally. In most instances, flocks did not return to a grazing area until a full year had passed. Today, most sheep producers keep their flocks on the same pasture year after year. If they use intensive grazing methods, the sheep also graze closer together. As a result, they must be wormed more frequently. Following is a list of the more common parasites and symptoms they cause. The list is divided into two parts, internal and external.

Internal Parasites
- *Roundworms*—When a sheep is infested with roundworms, it is often anemic. In severe cases, it will have a swelling under the jaw called bottle jaw. It may also become dull and listless or lose weight.
- *Lung Worms*—Lung worms are found in low lying, wet pastures. They live in sheep's air passages. Signs of lung worm infestation are coughing, accelerated breathing, listlessness, loss of appetite and nasal discharge.
- *Liver Flukes*—Adult liver flukes live in the bile ducts of sheep, cattle, horses, rabbits, deer, rats and people. They spend part of their life cycle attached to snails that inhabit wet areas. After 30 days, the flukes leave the snails and attach themselves to vegetation. Sheep eating this vegetation ingest the flukes.

    After 4-6 days in the intestines, the flukes move into the sheep's liver. Seven to ten weeks later, they migrate into the bile duct, where they mature and produce eggs. Symptoms of infestation are dullness, weakness and fatigue. Because liver flukes need snails to complete their life cycle, keep your sheep out of wet grazing areas (*The Sheepman's Production Handbook* green cover: 3-62).
- *Tape Worms*—Adult tape worms live in sheep's small intestines. Their bodies are made up of hundreds of segments, each containing egg-filled sacs. These segments break off and are passed out of the body in feces. Mites living in the pasture ingest the eggs. As sheep graze, they ingest the mites along with their forage. Tapeworms are then released into the small intestine and attach themselves to its wall.

    Unfortunately, tapeworms can infest sheep, dogs, and human beings. Sheep over a year old develop immunity to these worms and seldom have problems. Lambs intestines, however, can become blocked which often results in serious heath problems. Young dogs can get tapeworms by eating infected sheep carcasses or their byproducts (*The Sheepman's Production Handbook* green cover: 3-61).
- *Coccidia*—Small numbers of coccidial oocysts are found in most adult sheep. Because carriers generally exhibit no symptoms of infestation, they can deposit infected feces into your flock's water and feed sources without your knowledge. Stressed lambs are highly susceptible to coccidiosis. Lambs can

become stressed during transport, following a change in feed or as a result of over-crowding in feeding areas. Symptoms of coccidiosis are diarrhea, loss of appetite, dehydration, weight loss and weakness.

External Parasites
- *Nose Bots*—Nose bots are flies that deposit larvae at the edge of sheep's nostrils. The larvae then slowly make their way into the nasal passages and then into the sinuses. Nasal discharge signals infestation. Sheep will sneeze frequently. When nose bots attack, sheep will try to escape the flies by rubbing their noses against each other or on the ground (Simmons 1989:169-70).
- *Sheep Keds*—Sheep Keds, or Ticks, unlike other parasites, live their entire life on the backs of sheep. Female keds do not lay eggs. Instead, they incubate them inside their bodies until the maggots mature. When the maggots are fully developed, the female "glues" them on to the sheep's wool. Brown, protective cases soon form, allowing the maggots to develop into ticks. After 19 to 23 days, the ticks emerge. Adult keds only live about four days when separated from sheep. A female tick is capable of producing 10-20 larvae about once per week. When sheep keds are introduced to your flock, they spread rapidly (Simmons 1989:173-177).

    Ticks roam over a sheep's entire body, biting and causing skin irritations. Infested sheep often rub, scratch and bite their wool, breaking it off and reducing its quality. Fortunately, you can control ticks easily by dipping your sheep or spraying them with commercial insecticides immediately after shearing.
- *Lice*—Lice are the second most common parasite affecting sheep. Three types of lice exist, two suck blood and one bites. Eggs are attached to a sheep's wool and hatch in about 1-2 weeks. They develop for another 2-3 weeks before becoming adults. Infected sheep tend to scratch and rub themselves when infested with lice. Like ticks, lice are susceptible to insecticides. Generally, two treatments are needed (Simmons 1989: 178-9).
- *Wool Maggots*—Adult blow flies deposit eggs onto wet or soiled wool, which then hatch into maggots. These maggots feed on the wet wool. This causes open sores on the affected sheep. Check your sheep after a rain storm for signs of maggots. Sheep affected by wool maggots will lie in the shade with their heads and necks stretched out. They tend to generally start kicking with their hind legs if forced to move (*The Sheepmans Production Handbook* green cover: 3-67).
- *Scabies*—Scabies, (psoroptic mange) is caused by mites. Animals infected with scabies often rub their wool, leaving large scaly lesions. They also become anemic and thin. Although scabies is quite contagious, it can be treated by spraying your entire flock (at the same time) with a pesticide. If your sheep

become infected with scabies, you must report the infestation to state and federal authorities in all 50 states of the U.S. (Simmons 1989:178).

During late spring and summer, insects and parasites of all kinds can make sheep feel uncomfortable. When your sheep are on their feet, stamping their legs and shaking their heads, chances are insects are annoying them. Consequently, you should tend your flock daily to ensure your sheep are not feeling miserable. Even though it requires extra work, apply sprays and dips to protect your sheep.

- *Combat*—Sheep can be their own worst enemies. Late summer and early fall is the rutting season and, as Philip Keller relates, the time for battles among the rams for possession of the ewes. The rams proudly strut across the pastures and fight furiously for the ewes' favors. During this time, you can hear heads crashing and bodies colliding throughout the day and night. Shepherds know that some rams can injure, maim or even kill one another in this deadly combat. A simple solution exists; catch your rams and smear their heads and noses with axle grease. When the rams collide, the lubricant will cause them to glance off one another. Then they will stand around staring at each other stupidly. The grease dissipates much of the heat and tension with little damage done (Philip Keller 1970:122).

Good shepherding has positive consequences. Shepherds use their flocks to reclaim and fertilize farm land that has not seen animals for years. In Europe, farmers even hire shepherds to fertilize their fields. Nevertheless, many people think livestock (especially sheep) take from the land by consuming vegetation, cutting it and even ripping it out by the roots. In the 1800s, cattle and sheep ranchers on the American frontier fought over grazing land because cattle ranchers believed sheep left the grass so short, cattle could no longer graze on it.

Poor conservation of the grassland caused this problem, not the sheep. Grazing an area for too long will destroy its protective cover, regardless of the animals that graze it. Proper management can turn property from an eyesore into a park-like setting. Today, many ranchers graze sheep alongside cattle, thus increasing pasture yields and improving the land.

Most people picture a shepherd's life as peaceful and serene, the kind of life we all desire. Unfortunately, many people don't have time to tend a flock. For those fortunate enough to do so, the benefits are two-fold: the flock receives nourishment and good care and shepherds have the quiet times they need for their mental and spiritual well being.

Four

Wild Canines

"The pack appeared as tiny blurs moving across the ice. At first sight of the wolves, the herd seemed undisturbed; their great numbers gave them a false security. Big Grey and his four offspring were stalking slowly behind, only two hundred yards away when something startled one of the herd. Seeing their opportunity to divide the herd, the wolves made their rush toward the wedge between the herd and the panicked group. The group consisted of eleven caribou, three of which were calves. They were moving to the right towards the slope of the mountain. There waiting, was the white leader, until the caribou neared her position. She made her rush from the brush covered slope. Her onrush turned the caribou back toward the herd and the other wolves. Big Grey's strategy had worked. They now could keep the displaced caribou from returning to the herd and attack from two sides. As the pack charged the cluster, the caribou panicked, dispersing in all directions. One cow made the fatal mistake of turning into the charge of the white female. She lunged head first at the buckling cow, sinking her teeth in the shoulder of her prey. The male leader then struck from behind at the cow's upper flank, his force throwing the cow to the ice. As the four pups darted in and out nipping and barking with the excitement of the hunt, the cow regained her feet and began running. Big Grey was shaken loose and trampled. Unhurt and undaunted, he attacked again, this time ripping the tendons of her front leg. Stumbling again and falling, she would not again be able to rise. The black male had a grip on the cow's nose while the female leader ripped the jugular vein with quick shakes of mighty jaws. Then it was over. It had been a good kill" (Childs: 26-27).

In this chapter, we begin to switch our focus from sheep to the dogs that tend them. To understand the herding dogs of today, we first must understand their wild canine ancestors. Whether dogs descended from wolves or each evolved separately from a common ancestor,[17] wolves and dogs share common characteristics. In the wolf, however, these characteristics exist in a more primal form. Among dogs, they have become blurred through centuries of selective breeding. According to Kevin Behan, the process of canine domestication is "… nothing more than a remixing of primal patterns of wild behavior… "We've merely blended ancient balances of nature into new ratios to serve our modern needs" (Behan 1992:23). By studying the wolf and other wild canids, we can understand better the instincts and drives that are crucial in a herding dog's ability to cooperate with its human handler while moving livestock in a controlled manner.

Wolf packs usually range from four to seven members (although observations have documented packs of 30 or more). Whatever its size, for a pack of wolves to hunt successfully, a few requirements must be met: a system of hierarchy that organizes the pack; a system of communication permitting the maintenance of this hierarchy, and the drive to catch and kill prey.

Hierarchy

Wolf packs are highly structured societies and cannot survive if order disintegrates. All packs consist of an alpha pair and subordinate males and females. This order affects every aspect of pack life, including personality. Dominant wolves communicate their position by acting confident and outgoing while lower-ranking wolves act nervous and shy. Interestingly, when high-ranking wolves lose their rank, their personalities often change.

Hierarchy is communicated also by patterns of waste elimination and relates to breeding, feeding and hunting patterns, as well. Low-ranking male and female wolves urinate in a squatting position, which causes their urine to drill a narrow hole in the snow. Consequently, its scent is lost quickly. In contrast, high-ranking male wolves urinate with their legs raised, causing their urine to spread over a larger area. Thus, their scent is scattered more widely (Thomas 1995:28). High-ranking wolves also direct their urine at an object a higher percentage of the time than do lower-ranking wolves.

Wildlife biologist L. David Mech reported that leg-raised urination is twice as common on the perimeter of a territory as within it. Unfortunately, we don't know if or how this relates to dominance patterns because researchers have been unable to determine which wolves urinate on their territory's perimeter. Wolves also urinate at

---

17 "Recently, the American Society of Mammologists recommended that the domestic dog be reclassified as a new subspecies of wolf, *Canis lupus familiaris*" (Bush 1995:2).

trail junctions and about every 300 yards along trails (Mech 1991:53). Again, researchers don't have enough information to interpret these data with respect to dominance patterns.

For hierarchy to function efficiently, it must be communicated. But, how do wolves communicate without speech? The answer is simple; they communicate through body language. For example, wolves express themselves with their heads and tails. Generally, wolves bare their teeth in a snarl, point their ears forward and wrinkle their foreheads to express dominance (Mech 1970:82). They also close their jaws over the necks of lower-ranking wolves to remind them of their place in the hierarchy. Even though wolves have a crushing power in their jaws of roughly 1,500 lbs/in2 (double the strength of a German Shepherd dog) they use no more force than necessary to make their point (Lopez 1978:26). In contrast, wolves express subordination with a closed mouth, smooth forehead, slitted eyes and flattened ears close to the head (Mech 1970:26).

Status is also communicated in tail expression. A raised tail shows dominance, while a submissive wolf holds his tail low, tucked between his hind legs or curved forward along his hind legs. Diane Boyd, a well-known wolf researcher, can read the ranking of members in a wolf pack just by their tails: "(D)ominant wolves carried their tails high, while the subordinates carried their tails low; the lower the tail, the lower the rank of the individual" (Boyd in Steinhart 1995:13). Movement of the tail has various meanings. A loose, wagging tail indicates friendliness, while a quick, abrupt wagging of the tip or entire tail alerts another wolf to the possibility of aggression. Upon meeting a wolf of similar status, a high-ranking wolf will hold his tail high,

Figure 4.1 Facial expressions, adapted Shenkel, 1947 in Mech 1970:82

Figure 4.2 **Tail expressions, Shenkel, 1947 in Mech 1970:83**

while making it tremble. During mock, or play fights, attacking wolves beat their tails toward their opponent.

A she wolf is fertile for only 5-7 days in late winter or early spring. Unlike dogs, wolves are in estrus once a year and, according to Mech, the top-ranking pair, or alpha male and female, usually breed with one another. The entire pack, however, takes responsibility for the care and feeding of the pups (Mech 1991:89).

This pattern of reproduction is based on the hierarchical organization of the pack and has clear survival benefits. If two litters are born in a single year, the pups' chance of survival through the first winter is slim. When there are numerous pups, the pack can't provide enough food because packs rarely are successful in their hunting endeavors. Consequently, the pups become malnourished and weak. Pups must be almost full-grown, strong, and healthy before the onset of their first winter to survive. Most wolf packs can raise only one litter a year. If the pack includes a number of sexually mature members, what keeps them from mating, as well? The answer again lies in the hierarchal organization of the pack. Dominant female wolves can make subordinate females in heat sit down in the presence of males (thus precluding their ability to mate) using only body language e.g., the force of their stare (Thomas 1995:29). Similarly, the dominant male wolf sexually monopolizes the alpha female during her fertile cycle.

Wolves' social status is also established by play fighting. Wolf pups begin play fighting at approximately three weeks of age. Over time, this play fighting develops the hierarchical order within the litter, with the larger, tougher pups coming out on top (Mech 1970:69). When a wolf plays, it does so by grabbing an ear, neck or extremity. This occasionally causes pups to tumble over, leaving one on top and pinning the other to the ground. Wolf pups nip at each other's hind quarters as they play.

Survival Abilities

Each skill learned during play fighting enhances a pup's chance of surviving into an adult and helps the critical drives needed for survival: prey and defense. Observers have seen young pups chewing and tugging on soft objects, such as animal hide. At 69 days, they begin to pounce and chase after mice and other small animals, and at 75 days begin to rip and tear objects. (The tugging and pulling is the same action carried out when killing prey as an adult.) At 12-weeks old, they begin accompanying the pack on hunting trips. It is not until they reach 7-8 months, that they begin to hunt.

Wolves are able to concentrate on more than one task at a time, which is another skill critical to survival. Elizabeth Marshall Thomas "… once observed a wolf on a hill watching the progress of a caribou in the valley below. The wolf was near the den where her pups were playing but evidently she didn't want to take her eyes off the deer. Still, she felt that the pups also needed her attention so she slowly rotated one ear until it pointed in their direction. Thus she monitored them while continuing to watch the caribou …" (Thomas 1995:28).

Wolves possess extremely sharp hearing, smell and vision—much more acute than that of the domestic dog (Thomas 1995:27). According to Mech, wolves can hear sounds six miles away in the forest and up to ten miles away on open tundra (Mech 1991:26). Similarly, Peter Steinhart in his book *The Company of Wolves* tells about a woman who worked with a captive wolf pack located about a half an hour from her home. Each day, when she traveled to work, she arrived at a different time. "(B)ut whenever she arrived the owner was standing at the door, expecting her. He told her that the wolves had begun to howl fifteen minutes before she arrived, and they would howl in such a way that he knew it was she and not her husband, although they drove the same car" (Steinhart 1995:132).

Wolves' sense of smell is critically important for survival. It is extremely sensitive, over a hundred times more sensitive than that of human beings (Mech 1970:15). Mech observed wolves smelling a moose a mile and a quarter away (Mech 1991:26). Paul Joslin, of Wolf Haven, found that canines can discriminate odors in concentrations 100,000 times smaller than people can. (Joslin in Steinhart 1995:136).

Wolves have superior night vision compared to us. However, we have superior color vision. Biologists have shown dogs can distinguish blue from red but cannot distinguish either green from yellow or red from orange. Similarly, when red, green, blue and yellow dyes were placed on a base of clean snow in a captive wolf enclosure, the wolves most frequently detected the red and yellow stains (Bush 1995:30). Canines do not see in detail as we do; they see in general form. Their vision is highly sensitive to motion (ten times more so than ours), which is particularly useful in hunting. Wolves see first by movement, second by color and last by shape.

Hunting

Any wolf can initiate play interactions. However, alpha wolves are generally the most motivated members of the pack. That is why in intense situations, e.g., hunting or feeding, they are the leaders. Many people believe leadership in the hunt is simply a manifestation of the alpha wolf's dominance over other pack members. According to Kevin Behan, this distorts the nature of wolf leadership. Behan maintains alpha wolves lead, not because they physically dominate their pack mates, but because they are less inhibited. It is this lack of inhibition (in other words, their fearlessness, or better, their high level of confidence) that leads other pack members to key off of them. Subordinate wolves need alpha wolves to follow, otherwise, they cannot overcome their own inhibitions and remain relatively impotent in their actions (Behan1992:71).

According to Behan, each job is not so much a skill as a different emotional state of inhibition relative to rushing in on the prey. The more uninhibited a member is, the less sensitive to resistance he'll be, and the more direct in his drive to bite. He'll be the leader. The more inhibited an individual is, the more circumspect and restrained he'll act, and he will be a follower (Behan 1992:27).

To put it in human terms, alphas lead the hunt by virtue of charisma, not intimidation, which is not to say they don't use force in other contexts. They resemble John F. Kennedy more than your classic schoolyard bully. (This, of course, has important implications for handling skills, as we'll discuss later in the book.)

In describing wolf hunting behavior, Mech divides the hunt into five components: prey location, stalk, encounter, rush and chase stages (Mech 1970:194-203).
- *Prey Location*—Wolves use three methods to locate prey: direct scenting, chance encounter and tracking. Direct scenting is the most common method. To scent an animal, however, the predator must be directly down wind from it. Wolves can scent prey at great distances. Chance encounters occur when wolves sweep an area looking for prey. Wolves will sometimes follow fresh marks or tracks directly to their prey.
- *Stalking Stage*—Wolves approach their intended victim deliberately and cautiously, yet with their attention completely focused on the prey. The intent is to sneak upwind and as close to the prey as possible without alerting it.

- *Encounter Stage*—This is where wolves finally make contact. The prey can respond in three ways: first, it can stand its ground and defend itself; second, it can turn and attack the hunters; and finally, it can flee. Prey animals rarely respond by attacking. Most either stand their ground or flee from their attackers. Large animals like moose and musk-oxen tend to stand and defend themselves while deer tend to flee. If an animal chooses to stand its ground, wolves will often hold their ground, as well, and a stalemate results. Most wolves need their prey to run to stimulate the chase.
- *Rush Stage*—If the prey responds by fleeing, it triggers the wolf to bolt after it. According to Mech, this stage is the most critical. If wolves cannot get close enough, the prey escapes (Mech 1970:201).
- *Chase Stage*—This stage can be either short or long and often ends in failure. During the chase, the lead wolves control the direction of the hunt, while the others follow in single file. The chase usually does not last very long, for wolves will abandon the hunt if they do not catch their prey within the first two or three miles. If the pack does manage to get close enough, one wolf grabs the prey by the nose, while the rest attack it from the rear, grabbing hold of the animal's rump, flank or hind legs (Mech 1975:367-8). There is a belief that wolves kill by hamstringing[18] their prey. This is, in fact, unusual. If a wolf attempted to hamstring a moose, he would risk being killed by its hind legs. Thus, wolves avoid both the front and hind legs, attacking instead the rump, flanks, shoulders, neck and nose (Mech 1970:204).

Many people are familiar with *Never Cry Wolf* by Farley Mowatt. Mowatt claims that arctic wolves survive on mice. In reality, wolves more often hunt and kill large herbivores such as caribou, moose and deer. Hunting these animals is not an easy task and many times wolves are unsuccessful. For example, according to Steinhart, wolves approach roughly twelve deer or moose for every one they kill (Steinhart 1995:68). This represents only 8.3%.

Whether large or small, each kind of prey provides a different set of hunting scenarios and challenges. As a result, packs become specialists in hunting the particular prey most abundant in their area. Packs become so specialized that Mech questions if one accustomed to killing deer could survive a move to an area in which caribou or moose were more abundant (Mech 1970:205).

Moose will usually defend their calves throughout their first year of life. Nevertheless, calves are still wolves' first choice as prey. To kill calves, they must first separate them from their mothers. G. Atwell described how wolves accomplish this:

> "The wolf approached the pair of moose, whereupon the cow, with head lowered, charged the wolf which ran in an arc 60 to 75 feet from the calf. The cow

---

18 When a predator hamstrings an animal he cuts the Achilles tendon connecting the hock with the thigh muscles.

suddenly halted at the extremity of the arc (whereupon) the wolf ...attacked the calf. After 20-30 seconds the cow again rushed at the wolf which, upon evading the charge, returned to the calf" (Atwell 1964:313).

After 8 of these runs at the calf, the wolf prevailed.

While pursuing large herbivores, wolves generally trail behind their prey in single file. Individuals break away and run toward the front of the prey on its left or right side (in herding terms, flanking the prey.) When chasing prey, wolves often separate; some will follow the prey, while others run along its flanks. This can go on for miles until the prey becomes exhausted. Then the wolves attack (Mech 1970:211).

In the attack itself, some wolves grab the prey's hind leg just above the hoof, while others grab hold of its nose, as Childs described in this chapter's opening passage. A wolf's first choice, however, is to grab the rump because it is far away from both the front and hind legs. The other advantage is the rump is in the prey's blind spot. By tearing at its rump, a pack of wolves can stop the prey's forward motion.

Wolves use two methods to hunt caribou: ambushing and chasing to exhaustion. Coming upon a herd of unsuspecting caribou, the pack will sometimes divide. One wolf races ahead of the herd to a hidden location from where it can spring an ambush, while the rest continue pursuit from behind. The wolf lying in ambush will position itself uphill, giving it an advantage over the caribou. Usually, wolves cannot force their prey into an ambush so exhaustion is used as a weapon. One way to accomplish this is to run in relays. A pack divides into two groups. One group chases the prey, while the other waits behind. The task of the first group is to chase the herd back toward the second group. As they meet, the second group takes up the chase while the first group rests. This continues until the prey becomes exhausted and can be killed easily. Another method is to chase a large herd of caribou until a few fall back. These are then cut off and killed (Mech 1970:230-1).

Dingoes

When comparing hunting patterns of wild dogs with wolves, we find some interesting similarities and differences. Australian dingoes, like wolves, hunt both in packs and alone. Alpine dingoes hunt alone for small prey, e.g., birds and lizards, while open-country dingoes hunt in packs. When the prey is large, such as a baby kangaroo or calf, dingoes must hunt in packs to be successful. Dingo packs are three times more successful in "bailing up" or confining kangaroos, than single dingoes. The lead dingo drives the prey toward the rest of the pack which then cuts off the animal's escape route (Corbett 1995:113). Sometimes dingoes will harass their prey for days. Members of the pack alternate between attacking mothers and their calves and resting. The mothers become exhausted and leave their calves to find food and water. Often, the calves are dead before their mothers leave (Corbett 1995:116-7).

In hunting cattle, dingoes often take advantage of sparse watering holes. Because of the distance between wells, cattle tend to become confused and congested at these water troughs. The dingoes' attack spooks the herd, causing a stampede in which mothers and their calves sometimes become separated. When this occurs, the calves become easy targets. Once their prey is separated and "bailed up," dingoes nip or hamstring the animal. This slows the prey and the dingoes either attack its throat or run alongside biting its neck and ribs.

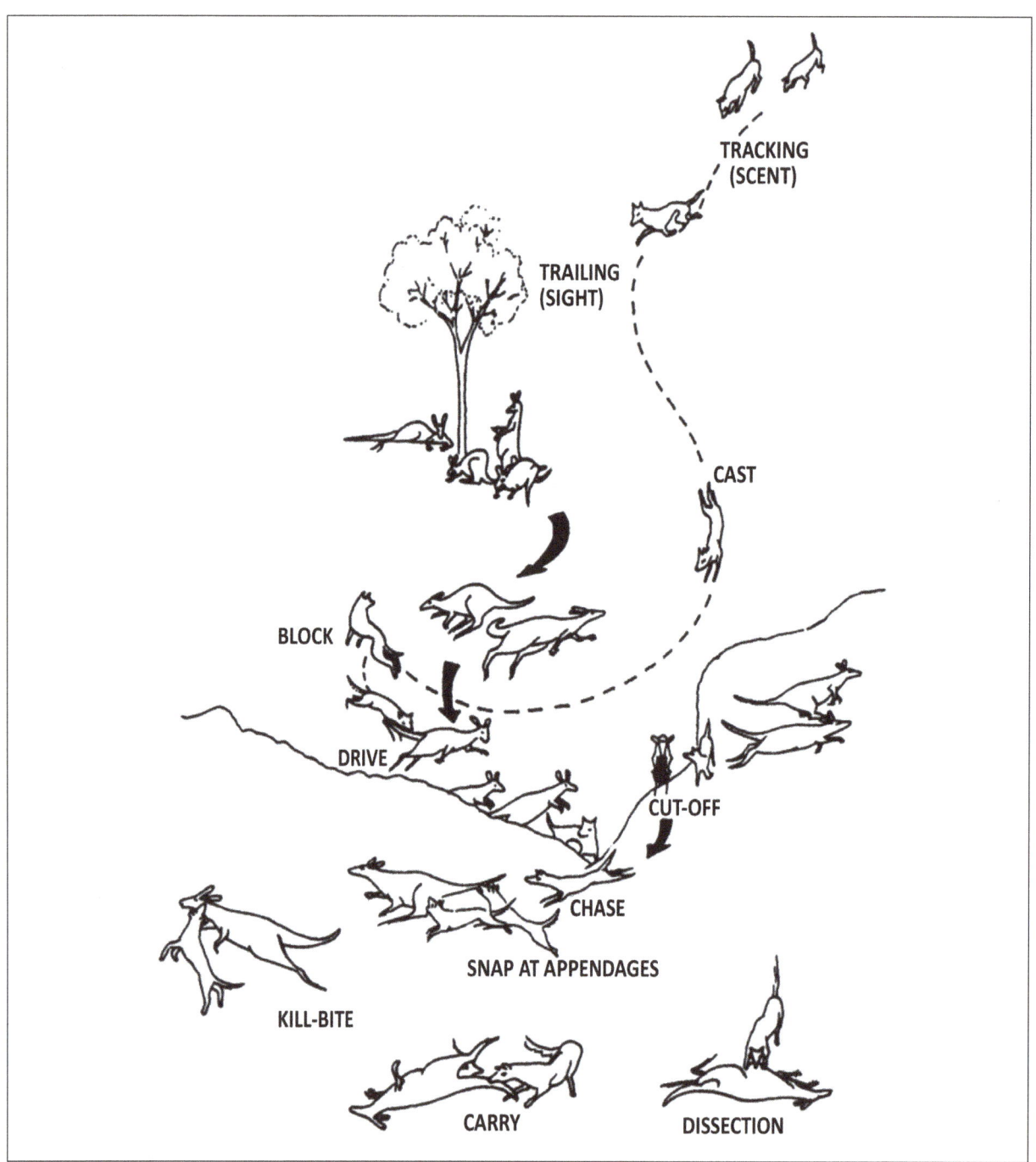

Figure 4.3 **From Morris 1987**

Kangaroos can sense dingoes' presence at 121-150 and flee from attack at 98-105 meters. This affects dingo hunting techniques. In his diagram of a typical dingo hunting sequence (see Figure 4.3.) Don Morris identifies six basic elements: tracking, cast, block, drive, cut-off, and kill-bite (Morris 1987:198-200). Because kangaroos are so wary, dingoes must approach their prey carefully. This necessitates a cast much wider than the flanking maneuver of wolves. It is more similar to the outrun of Border Collies. Once contact is made, dingoes block avenues of escape and drive their prey in an advantageous direction. Finally, the dingoes are able to cut-off a single animal and begin the final kill.

Dholes

Like dingoes, dholes (a kind of wild dog found in India) hunt alone and in packs. As with wolves and dingoes, their hunting techniques adjust to differences in type of prey and hunting conditions. Unlike wolves and dingoes, these dogs eat their prey alive. The dholes grab their prey along its flanks and between its hind legs or thighs, tearing out large chunks of flesh that are then consumed immediately. As the prey weakens, the dogs circle and attack. By circling quickly and changing directions, two dholes can give the impression they are an entire pack. This confuses their prey, making it easier to kill them (Davidar 1975:109-119).

Domestic Dogs

How do domestic dogs differ from their wild canine relatives? Dogs are born with similar instincts and drives. Because we have close relationships with our dogs, they become so much a part of our lives; we forget they have a wild side to their nature. Domestication has added nothing new to the basic patterns of canine behavior. We merely have remixed the old combinations to suit our own needs (Behan 1992:23).

Feral Dogs

To see more clearly the effect domestication has had on wild canids, study feral dogs. The process of canine domestication occurred over tens of thousands of years. Yet, we often allow our dogs to roam free, outside our control, to become wild again. Although these free-roaming pets form packs, their members rarely act as a cohesive unit under the direction of one leader. They are merely groups of loosely organized individuals whose membership fluctuates. Packs of domestic dogs sometimes roam the countryside killing livestock. It has been determined that dogs caught killing sheep were raised as herding dogs and even guard dogs. In the novel *Bob, Son of Battle*, written in 1898, the author describes the killing exploits of Red Wull, the dog villain in the book.

If had bin ony other dog-greyhound, bull, tarrier, or even a young sheep-dog—d'yo' think he'd ha' stopped wi' the one? Not he; he'd ha' gone through 'em, and be runnin' 'em as like as not yet, nippin' 'em, pullin' 'em down, till he'd maybe killed the half. But 'im as did this killed for blood, I say. He got it—killed just the one, and nary touched the others, d'yo' see, Jim?'

… An owd dog'll git the cravin' for sheep blood on him, just the same as a mon does for the drink; he creeps oot o' nights, gallops afar, hunts his sheep, downs 'er, and satisfies the cravin'. And he nary kills but the one, they say, for he knows the vallie o' sheep same as you and me. He has his gallop, quenches the thirst, and then he's for home, maybe a score mile away, and no one the wiser i' th' mornin'. And so on, till he comes to a bloody death, the murderin' traitor (Ollivant 1898: 146).

During the day, Red Wull was one of the best herding dogs in the community, a trusted helpmate obeying his master's commands. But at night, when he was left to run alone over the hills and through the marshes, he became the "black killer." It is interesting to note that he did not attack his own sheep. He only killed the sheep belonging to a neighbor.

As dogs wander, they sometimes hear the beckoning call of the wild and their human ties dissolve. When they are completely broken, the dog becomes feral. Today, ten percent of the world dog population is feral. In 1983, in Italy alone, there were approximately 3.5 million dogs. Of these, 850,000 were free-ranging, among which were about 80,000 feral dogs that lived in the wild, avoiding contact with people (Ciucci and Botitani 1989:10).

Even feral dogs need companionship and, like their wild cousins, form packs. However, these packs cannot organize their hunting activities effectively because domestication has inhibited their ability to kill. In *The Dog, Its Domestication and Behavior*, Michael Fox describes the effects of domestication on a dog's ability to catch and kill prey. He based his findings on observing 4 beagles, 9 beagle/coyote hybrids and 13 coyotes age 8-9 weeks. The prey-catch-kill sequence among wild canids is normally:

1. Approach
2. Investigate
3. Bite
4. Carry
5. Bite (and head-shake) to kill
6. Dissect
7. Ingestion

(Fox 1978:112)

The beagle sequence of prey-catch-kill, was cut off at the bite phase. The beagle-coyote hybrids were inhibited in their biting, like the beagles, but ate their prey before killing it. Fox concluded that domestication affects dogs' ability to complete the prey-catch-kill sequence. He also showed how the sequence varies among different breeds of dogs. Figure 4.4 illustrates the complete sequence of predatory behavior of wild canids. The dotted horizontal lines mark cut-off points or partial sequences of some domestic dogs. For example, sheep dogs track, herd and drive, but they do not stalk, point, attack, kill or retrieve.

The traits of the Border Collie, however, are not cut off at the herding, driving stage. They flow past into the stalking, pointing stage. According to Fox, this breed should not exhibit the traits of the Setter or Pointer. However, Viv Billingham, a well-known Scottish Border Collie handler and breeder, believes our ancestral shepherds' collies were crossed with Setters. Some believe this produced the Border Collie and gave the breed the eye and stalking traits we see today.

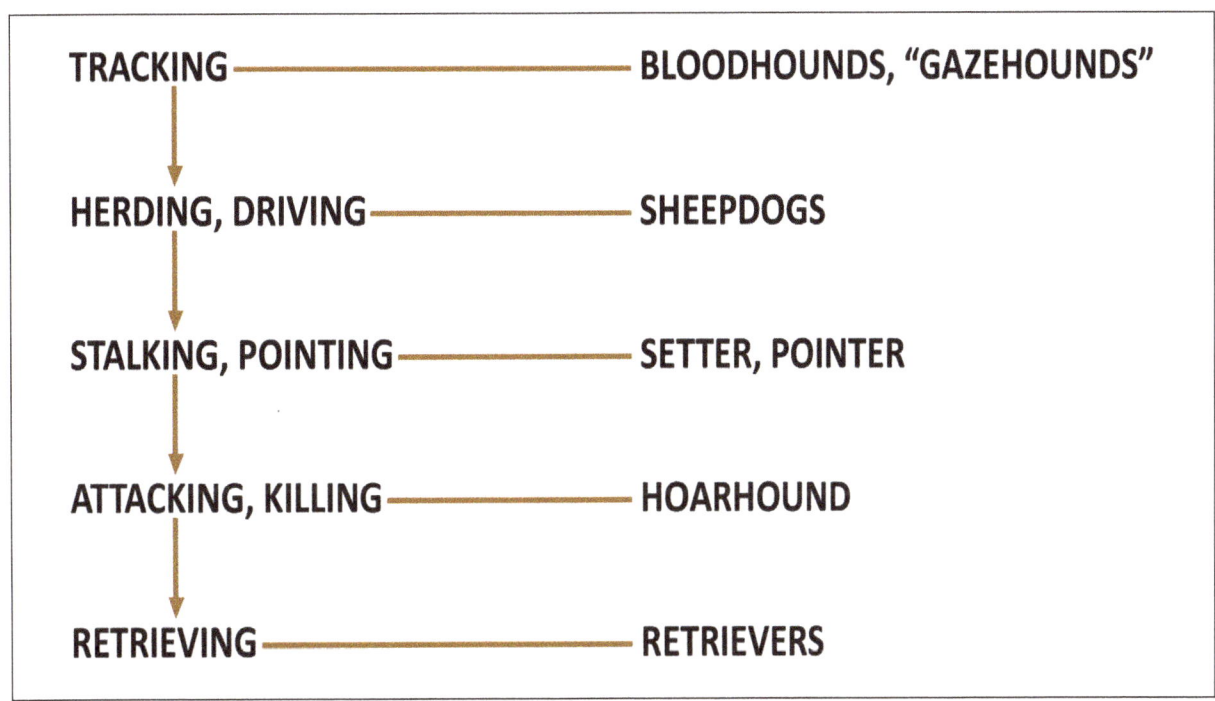

Figure 4.4 Sequence of predatory behavior of wild canids, adapted from Fox 1978

Neoteny

How has the instinct to kill been shaped and suppressed in the process of domestication? One way was through neoteny—the retention of juvenile traits in the adult members of a species. Psychologist Harry Frank concluded that domestication selects, among other things, infantile qualities such as, docility and dependence. For example, when dogs are placed in pens with wolves, they are not killed. On the contrary, wolves treat the dogs as puppies (Steinhart 1995:129).

Biologists Ray and Lorna Coppinger have categorized dogs by the degree of neoteny in various breeds. They identify four developmental stages in the maturation of wild canids. Stage one is characterized by the typical patterns found among both wild and domesticated canines. Pups do not stray far from the whelping box (in the wild, from the den). Mothers first nurse their pups and then feed them regurgitated food. Both wild and domesticated pups are very responsive to stimuli and, at the slightest movement, run back to the den or whelping box for safety. In the next stage, wild pups begin to play with different objects, such as, each others' tails, bugs or sticks. During stage three, the pups lie in wait, stalk and pounce on unsuspecting objects, i.e., a stick, leaf or even another pup. Motion is usually what triggers these responses; the responses themselves are aimed at stopping the motion. In the fourth stage, pups tag behind the adults during the hunt. Their focus changes from stopping motion to creating it. Only by combining the instincts to stop and start motion, can wild canids hunt their prey successfully.

According to the Coppingers, different breeds become retarded at different stages of wild canine development (Coppinger 1982:71). Figure 4.5 arranges breeds according to the degree of juvenile characteristics retained. At the top are breeds retaining the

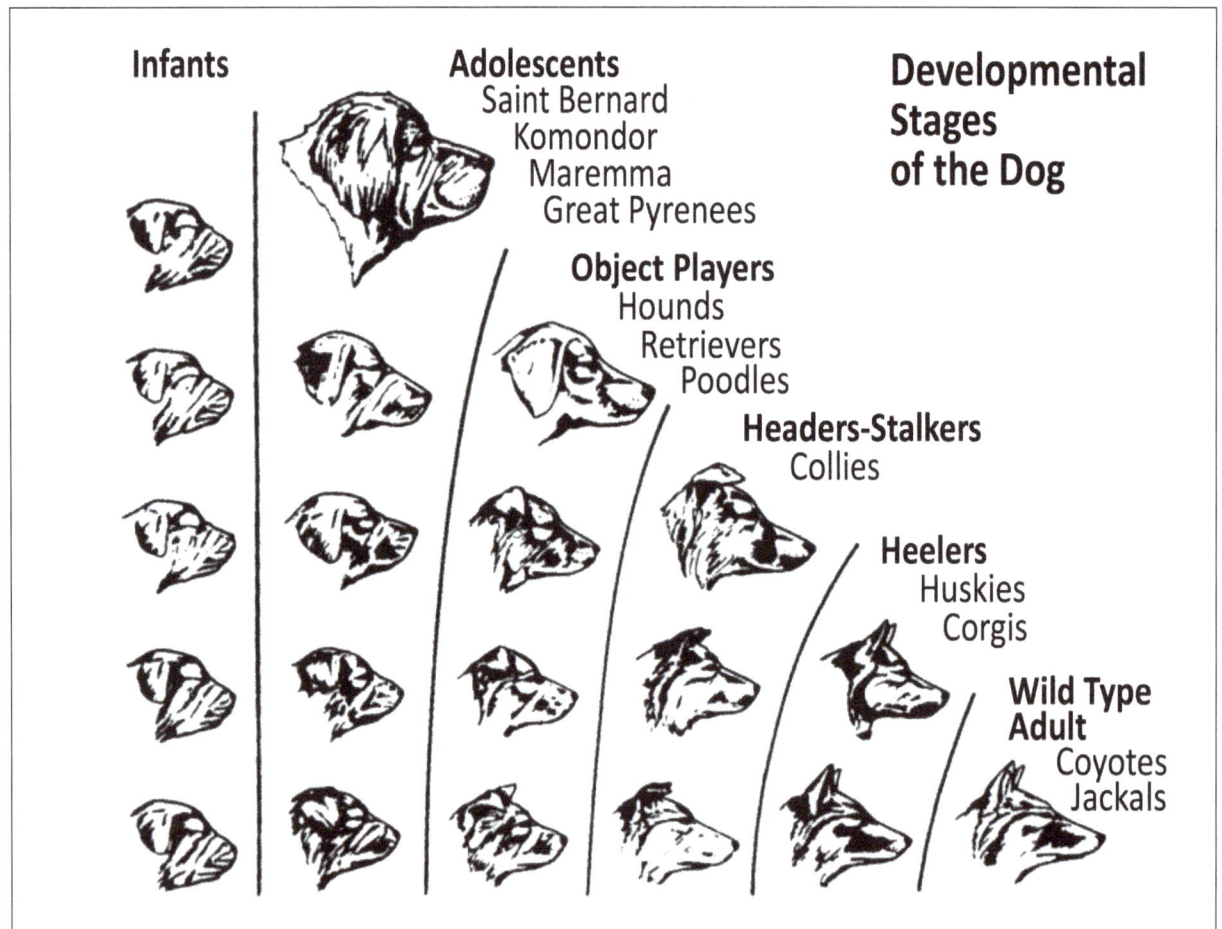

Figure 4.5 **Adapted from Developmental Stages of the Dog, Coppinger 1982**

most juvenile characteristics and at the bottom are those retaining the least, meaning they most closely resemble wild canids when they are adults.

Domestication dampened the instinct to hunt and kill prey and made dogs more docile and dependent though the process of neoteny. Without it, our dogs would be untrainable. They did retain many other traits from their wild canine ancestors and some of these became both the foundation for, and the limit of, our dogs' ability to work with human handlers in controlling livestock. If we are to utilize the capabilities our dogs possess, we need to identify the basic behaviors of the hunting instinct they inherited.

What have we learned about our dogs' herding capabilities from studying wild canine hunting? As did their ancestors, herding dogs possess the instinct to start and stop motion, which forms the foundation of herding behavior. Once they start motion, they can control its direction by using their sense of balance, selective flanking, and moving to the head and biting a nose, if necessary. They can drive stock away from a particular point or cast out and around to push stock toward it. They can stop motion already started by moving to the head and blocking the direction of flow.

Once motion is stopped, they can keep stock contained by continual circling and by the force of judicious, controlled grips. They can separate designated animals from the herd. Finally, and most importantly, they can work as part of a team, taking directions by keying off their leader. As we shall see in later chapters, training herding dogs to tend builds on these abilities, shaping them in particular ways to suit our needs.

Intelligence

Domesticated dogs differ from their wild ancestors in a way we have not discussed, a difference that is critical for herding. Pack leaders do not give orders to subordinates during the hunt. A wolf pack coordinates its behavior during the hunt through a keying process, i.e., subordinates key their movement off of one another, their prey, and most importantly, the pack leader. Understanding this has clear implications for developing our own handler skills. Without our dogs' ability to follow orders, herding would be impossible. This takes us to the nature of dog intelligence and how it is both similar to, and different from, the intelligence of wild canines.

In his book, *The Intelligence of Dogs*, Stanley Coren describes three types of intelligence: adaptive, working and instinctive. Adaptive intelligence is the ability to learn and to solve problems. Simply stated, it is the ability to piece together bits of information to form a solution or to discover new ways to apply previously learned information. It is measured by the number of times an animal has to experience something to understand it.

Wolves have a great deal of adaptive intelligence. For example, it is very difficult to trap a wolf twice. One experience is enough for it to learn not to go near traps. Because of this ability, the 3M Company developed a radio controlled collar that can dart a wolf on command. Wolves need to be trapped only once to put on the collar. After that, they can be recaptured with darts.

In *The Company of Wolves*, Steinhart describes three canines (a wolf, a wolf hybrid and a Malamute) belonging to a psychology professor at the University of Michigan. During six years of watching people go in and out of the kennel area, the Malamute never could unlatch the door.[19] The hybrid learned to unlatch the door after watching for only two weeks but, the wolf's learning capability was even more amazing. After watching the hybrid unlatch the door only once, the wolf could unlatch it himself. Moreover, the wolf used a new technique to open the door. While the hybrid used his muzzle, the wolf used his paws. Wolves seem to study things and think through problems internally. Even wolf puppies study objects to figure out how best to manipulate them (Frank in Steinhart 1995:128-129). In contrast, dogs solve puzzles only after extensive pawing, tugging and prodding, i.e., by going through a long process of trial and error (Steinhart 1995:129).

Working intelligence is the ability to follow orders or take direction from a leader; it does not necessarily go hand in hand with adaptive intelligence. For example, a student of Harry Frank, a well-known animal behaviorist, tried to teach a wolf to respond to two basic obedience commands: sit and heel. He worked with the wolf for six months, but it never learned the commands. A wolf hybrid watched a dog heel and sit on command for six days and executed the commands as well as the dog did. This led Frank to set up more experiments comparing dogs' and wolves' intelligence. He found that, while dogs are good at learning commands, they are not very good at solving problems. In contrast, wolves don't learn commands very well, but are very good at problem solving (Frank in Steinhart 1995:129-131).

Some herding dogs possess both strong working and adaptive intelligence. Others, however, follow commands well, but cannot think on their own. Some dogs can be trained to follow commands to accomplish a given task, but are unable to accomplish the same task when out of range of their handler's signals.

Many Border Collie herding competitions test adaptive intelligence with "the silent gather." After a single command, the dog must complete an outrun,[20] lift,[21] and fetch[22] for a group of sheep several hundred yards away. The dog must bring the sheep in a straight line to the handler. The handler may not give the dog commands to help it fetch the flock. Instead, the dog must rely on its adaptive intelligence. Only the truly

---

19 Unlatching the door required two different operations. First, the handle had to be pushed toward the door and then it had to be rotated.
20 The dog runs from the handler's side (to left or right) to reach the far side of the sheep.
21 The moment when the dog has reached the end of its outrun and moves forward to initiate contact with the sheep.
22 The dog drives the sheep toward the handler after lifting them. On a trial course, the fetch includes the movement of the sheep in a straight line.

great dogs that meld wild and domestic intelligence capabilities do well in this type of competition.

Instinctive intelligence is comprised of genetically based behaviors and skills passed on from generation to generation through selective breeding. For instance, if you want a dog that possesses a sense of balance,[23] you have to breed for this. Balance is one of those "… genetically determined abilities and behavioral predispositions… that can be transmitted from generation to generation through the biological mechanisms of inheritance" (Coren 1994:125).

Instincts and Drives

Instinctive intelligence is only one genetically based element we need to understand in training our dogs. We also need to understand general instincts and drives. We'll begin with the difference between instinct and drive. Did you ever ask yourself what motivates a dog to kill a wounded animal? What provokes it to chase a cat? What causes a mother to seek out food for her puppies? What makes a dog defend itself when attacked? What drives a young dog to run with a pack? What induces a puppy to toss a stick into the air? The answer to all these questions is drive: territorial, prey, food, defense, pack, play and sex.

Drives are related to instincts, but combine, develop and focus them in specific ways. As Sheila Booth explains:

> "Most drives, unlike instincts, can be built or heightened if they are present. Conversely, they can also be suppressed and often even extinguished, unlike instincts. Thus it is usually easier to get a dog to stop retrieving or herding than it is to stop him licking, scratching or digging.
>
> A simple way of understanding the difference between the two is that the dog's drive determines the degree to which he exhibits instinctive behavior. So while it is instinctive for a dog to salivate and eat, food drive determines how often, how eagerly and how intensely he pursues food." (Booth 1992:xvii).

Thus, drives motivate our dogs to act in the ways they do. While salivation is an instinctive behavior, the desire for food drives our dogs to salivate.

A dog's daily activities revolve around its various drives and when these are set in motion, different reactions occur. For example, Spot wakes up in the morning and is promptly let outside. As he patrols the boundaries of his domain, he marks his territory by urinating (communication and territorial drives). Then he returns to the house to eat (food drive). To eat, he first must defend his food from the cat. He growls and snaps at her, warning her away from his food (defense drive). She retreats only to find

---

23 The distance a dog needs to be from the sheep to maintain contact and control without upsetting or scattering them.

a ball she can roll past him. As she rolls it by, Spot stops eating, runs after it and begins tossing it into the air (play drive). While all dogs possess the same drives, they do not operate in exactly the same manner in all dogs. Suppose during a herding instinct test, one dog mounts a sheep to breed it, another bows in front of it to play, while a third circles it before driving it into a corner of the pen. All three dogs were confronted with the same stimulus, however the stimulus triggered behavior motivated by the sex drive in the first, the play drive in the second and the prey drive in the third.

The history of the wolf's domestication is a process of shaping canine drives so that its dog descendants became helpmates in meeting the subsistence needs of our ancestors. Herding dogs, in particular, were created by shaping and modifying the wolf's pack drive, prey drive and defense drive. The instinct to control stock and the willingness to work with a human handler we see in herding breeds today is a direct result of this process. The key to working a herding dog is to use its drives to our advantage. As you observe your dog, you will see it respond to particular situations in different ways. Your job as a trainer is to use this information to instill the behavior necessary to complete the herding task at hand.

Herding training mainly involves utilizing the motivations of the prey, defense and pack drives. Prey drive involves the instinct to chase anything that is in motion and kill it (although the killing part has been severely dampened, if not extinguished, in the process of domestication). The defense drive involves both fight and flight. When an animal is confronted, it will react in one of two ways: stand and fight or run away. We use the dog's defense drive to confront stock and stop their motion. When your dog submits to your authority by obeying your commands, you are utilizing the pack drive.

We can use play and food drives to our advantage, as well. The play drive takes center stage when your dog entertains itself, while the food drive emerges when your dog is hungry. Our goal is not to shape the behavior motivated by these drives. On the contrary, it is to use their motivational power to shape behavior directly related to herding. Successful training involves shaping certain aspects of our dogs' drives by using other aspects as rewards to get the behavior we want. We use drives whenever possible to shape, enhance or dampen other drives.

All breeds possess the same drives, but they possess them to different degrees. The prey drive is usually the strongest in herding and hunting breeds. This is not to say that other breeds do not exhibit prey drive, as well. Within hunting and herding breeds, individual dogs may have stronger defense, food, play, pack or sex drives. One dog I trained had an exceptionally strong defense drive. He would only herd the livestock with authority if a sheep or cow challenged him. Otherwise, he showed no sustained interest in herding. Another possessed a strong pack drive. For months, she refused to leave her owner's side. Fortunately, we were able to enhance and shape the dog's prey drive and ultimately, she became a useful herding dog. Finally, I worked with one dog that displayed an exceptionally strong sex drive. He was only interested in mounting sheep. Although he was a Border Collie, a breed known for its herding capabilities, this particular dog never could focus on herding livestock and I had to abandon his training.

Prey Drive—Herding and Hunting

What separates herding from hunting dogs? It is not uncommon to observe a hunting dog with herding behaviors or a herding dog with hunting instincts. Herding dogs control livestock utilizing their prey instincts. Herding livestock is just a specialized form of hunting.

When wolves hunt as a pack, they make use of their members' strengths. One wolf may be good at tracking and trailing, another good at casting and blocking, a third an expert at cutting off and driving and a fourth good at chasing and killing prey. The same applies to hunting and herding dogs. Thousands of years ago, our ancestors who wanted a hunting dog placed their breeding emphasis on tracking, trailing, carrying and retrieving. In contrast, those who wanted a herding helpmate focused on developing smell, sight, casting, blocking and driving abilities.[24] Let us look at each of these in detail.

Most herding dogs do not seemingly use scent to find sheep (although they use it to identify intruders). A tending dog does use his sense of smell to track back and forth along a boundary. It also uses its keen vision to watch over its flock. If a sheep crosses the boundary of its assigned grazing area, the tending dog immediately moves to force the sheep back into the flock. The tending dog also uses its eyes to keep in contact with its master.

Many herding dogs, like the Border Collie, are known for their wide cast or outrun as they approach livestock. Tending dogs, however, work much closer to the stock. Consequently, their cast is shorter and narrower—they only cast far enough around the sheep to force them into their grazing area. Patrolling around the grazing flock is a variation of casting.

All herding dogs use blocking as a technique to control livestock. Large cattle dogs, such as Bouvier des Flanders and Rottweilers, use their bodies to block and shoulder a cow, to make it change directions. The block is an especially important tool for tending dogs. They must block or turn sheep's paths back into a grazing area or onto a roadway.

In general, herding dogs should not use their ability to cut off individual members from the rest of the flock. Their job is to keep the flock together. This is because once an animal is effectively cut off from the flock, dogs quickly move into a chase mode

---

24 I use "driving" as a general term for the controlled movement of stock. Sometimes, dogs move stock toward and sometimes away from their handler. In any case, all driving is a shaped version of wild canine hunting behavior.

and the joy of killing rises to the surface. On rare occasions, an animal must be separated from the rest of the flock. Only under a handler's command should a herding dog separate one or more animal from a flock.

Finally, herding dogs snap at appendages to move and punish livestock. Fetching/driving dogs nip at an animal's head or heels to get it moving, changing its direction or stopping its motion. Tending dogs grasp and hold (without tearing flesh) an errant sheep on its neck, ribs and upper thigh (See Gripping, Chapter 9).

Defense Drive

A fine line exits between the fight and flight aspects of the defense drive, although it is important that the fight part of the drive is strong. Herding dogs must stand their ground when challenged. If they back down from opposition, the livestock will take it as a sign of weakness and resist their pressure. In some instances, just turning their head away from the livestock signals weakness. If a dog wants to flee, rather than stand its ground when confronted by strong-willed livestock, it should be trained in a small area, for it will force the dog to defend itself. If the dog still cannot stand its ground, then it is not suited for herding work. On the other hand, the dog that wants to fight too much of the time must be taught to control its defense drive. To do this, the dog must be taught to bite only on command.

Photograph 4.1 Belgian Sheepdog in defense drive

Food Drive

When dogs are hungry, they move into the food drive and the hungrier they are, the stronger the motivational power of the drive. You can use this to your advantage teaching the meaning of commands. If your dog receives food during its training sessions, however, it will not need as much food in its feed bowl. You must be careful not to starve your dog, for this obviously is dangerous for its health. If your dog wants food and nothing else, it will be difficult for you to use other drives for its training. A dog's need for food triggers and enhances the prey drive. This can be advantageous or disadvantageous, depending on the natural strength of your dog's prey drive and your training goals.

Play Drive

Most dogs love to play. Playing sends their spirits soaring, boosting their attitude about life, in general. For this reason, many obedience and Schutzhund trainers incorporate a dog's play drive in their training exercises. If you reward your dog with play only after the training session is over, it will not learn that training itself is fun. Consequently, it will not approach training sessions with the same enthusiasm it would otherwise. The key to using the play drive successfully is to incorporate it into your training, in addition to using it as a release when training sessions are over.

Although play is widely used in obedience and Schutzhund training, many herding trainers disapprove of its use in training herding dogs. Nevertheless, it can be used productively in both off-stock and on-stock work. It is easy to understand its usefulness in off-stock training because it is similar to competition obedience training. Its usefulness in on-stock work, however, is less obvious. Still, it does have its place.

Photograph 4.2 **Briard in play drive.**

For example, your dog may need to move into play drive to boast its attitude after a discouraging experience. A dog that has been challenged by a sheep and has lost the battle needs its spirits immediately lifted. By transporting it into the play drive, you can get it to forget about the discouraging situation. Otherwise, your dog's refusal to continue working and, its last experience with stock during that session will be embedded firmly in his memory. This will make the next session more difficult.

On the other hand, your dog must be in a controlled prey drive (not a play drive) while herding or it will not accomplish useful work. A dog can move very quickly from play to an uncontrolled prey drive, resulting in damage to the livestock. Be very careful your dog does not remain in play drive for too long. You can tell if your dog is playing if it: chases the stock without purpose; bows and bounces in front of them or runs with its tail up in the air. Observe your dog's body language carefully as it herds to ensure it is in the prey drive.

Good Herding Dogs

(handwritten: Five)

Characteristics of a Good Dog

A good herding dog must possess the following qualities:
1. Prey Drive
2. Pack Drive / Working Intelligence
3. Adaptive Intelligence
4. Sustained Interest

Throughout chapter 4, we learned about the importance of prey drive in the life of the wolf pack. It is this drive that also motivates the herding dog to control livestock. As von Stephanitz noted:

> "This instinct always to circle round the herds was an excellent quality, making them ideal for work with the flocks and herds, because it made it easier for the shepherds to keep their charges together both on the road and on the pasture" (von Stephanitz 1925:69).

We also discussed the importance of the pack drive. Without a willingness to work together and submit to the alpha pair, the pack cannot survive. John Holmes, in his book, *The Farmer's Dog*, describes a fine balance between what he calls the "herding instinct" and the "submissive instinct" (what we call the prey drive and the pack drive).

> "The herding and the submissive instincts are among the most important factors which combine to produce a good working dog. Most important point of all is that the one should as nearly as possible balance the other. The herding instinct makes the dog want to herd, to use a common phrase, "start to run." The submissive instinct gives us what we call a biddable dog, which can be

kept under control. The stronger the herding instinct, the more necessary it becomes to have an equally strong submissive instinct to balance it." (Holmes 1984:27)

Similarly, working intelligence is closely connected to the pack drive. A dog must be willing to submit to its master, but it also must have the ability to recognize and respond to its master's signals.

It is adaptive intelligence, however, that separates a good working dog from a great one. A dog with the ability to think for itself can work much more effectively and can adjust to new situations faster than its handler can. For example, many herding dogs can move stock, but few can move them through obstacles unless the handler gives them numerous commands. If your dog can adjust to the obstacles in its path without help from you, it has adaptive intelligence.

The final characteristic of a good herding dog is sustained interest. Many dogs possess all of the qualities discussed, but lack sustained interest. They work for a while, but soon become more interested in other things, a cat, squirrel, etc. Without sustained interest, the other qualities are negated, for such a dog cannot see a task through to its completion.

Unfortunately, you may have to train several puppies before you find the one that possesses all these characteristics. Do not settle for a couple of these vital qualities; continue your search until you find a dog with them all. Once you have worked with a dog of this caliber, you will never be satisfied with a lesser quality animal.

Choosing a Good Herding Dog

I once got a phone call from a man with a serious predator problem. Every night, coyotes attacked his sheep and goats and killed some of them. He talked to a breeder who advised him to buy a Rottweiler, put it in the pasture and the problem would be solved. Rottweilers, according to the breeder, are instinctively protective. The man took the advice. Unfortunately, the dog moved the sheep and goats from one part of the pasture to another for hours, never giving them rest or time to eat or drink. Obviously, the Rottweiler was not the solution. He then talked to another person who told him Border Collies are the greatest sheepdogs in the world. He called me to purchase one of my dogs. This man needed a Border Collie no more than he needed a Rottweiler. What this man needed was a livestock guarding dog (Akbash, Great Pyrenees, or Maremma-Abruzze Sheepdog), a dog bred to guard stock against wild predators.

The point of this story is simple. In choosing a herding dog (or any dog, for that matter) it is important to assess your needs carefully and thoroughly research your options. To help you begin the search, I have compiled a list of livestock dogs. This list categorizes dogs by the kind of work done, i.e., herding/guarding, and then by the kind of livestock herded. It also includes a brief description of sheepdog breeds (see Appendix III). I have excluded British, Australian and New Zealand breeds. You will not find information on, for example, Border Collies, Kelpies, English Shepherds or Huntaways, only because this book is about training dogs to herd in the European tending style. I have included only those breeds originating in that part of the world.

Selecting a herding dog is much like planting a garden. Because there are so many choices, the task of choosing what to plant seems daunting. To make it less overwhelming, make your selection in as logical a manner as possible. One of your first decisions is whether to plant vegetables or flowers. If you need to feed your family, then a flower garden will do you no good. Similarly, your first decision in choosing a stock dog is whether to get a dog bred to guard a flock or to control sheep.

Let's assume you need a dog that works in the tending fashion. Your weeding-out task is not finished. Do you want a dog you can send from your feet, but that returns to you when its job is completed? Do you want a dog that once sent, works at a distance? For instance, Pulis, Schapendoes, Old Fashioned German Shepherds and Polish Lowland Sheepdogs work close at hand. In contrast, German Shepherds, Briards, Dutch or Belgian Shepherd Dogs and Beaucerons are bred to work at a distance, independent from their handlers.

Take a closer look at the breeds that fit into working style categories. This is much like the peppers you grow in your vegetable garden. They are either sweet or hot. They also come in different colors: red, green, yellow, orange and deep purple. What you choose to plant comes down to personal taste. Similarly, once you choose a herding style, you can choose a dog based on appearance, i.e., size, color and coat. For example, if you want a medium size dog, you might choose a Dutch or Belgian Shepherd. If you want a larger dog, a Beauceron, Briard or German Shepherd would be a good choice. If you like a dog with a corded coat look at a Puli, Spanish Water Dog or a Bergamasco. When considering the type of coat you want, keep in mind the terrain on which the dog must work. Many dogs were bred with coats to protect them from rough terrain. A corded coat provides a dog with protection from harsh weather conditions and the teeth of attacking predators. Keep in mind too, that long-coated dogs, e.g., Briards or Shapendoes, should have goat-like or harsh, rather than soft coats. Soft coats pick up debris: twigs, seeds, brambles, quickly. If you like everything about a dog except its long or corded coat, you can cut it off once a year when you shear your sheep.

Once you have decided on a particular breed, it is time to select a puppy. The best place to start is with a kennel club. (A list of kennel clubs and their addressees is included in Appendix IV.) They can give you a list of breed club secretaries who can give you detailed information about their breeds and help you find reputable breeders in your area. You should contact several breeders. Ask them if their dogs are bred for herding ability. Just because a dog is categorized as a herding breed does not mean that it was bred for herding abilities. On the contrary, many dogs from herding breeds are bred solely for their physical characteristics.

Arrange a time with the respective breeder to observe both the sire and dam working livestock. Ask to see several of their offspring, as well. If none are present, ask for a list of owners whom you can contact. If they live far away, see if a video of their dogs is available. Also ask for a list of people who have bought puppies from other parents owned by the breeder. Call these buyers to ask how well their pups herd and if they found this breeder to be professional. Also ask if they have had any health problems with their dogs. Look at several litters. Too many prospective buyers purchase a puppy from the first breeder they contact. The process of buying a pup should be similar to buying a car or choosing a college. Remember, you are buying this dog for a purpose, to be an effective helpmate that tends your flock. You want to buy the best dog possible. Time is on your side, so research your options thoroughly before you choose.

Also ask the breeders how they raise their puppies. Puppies' brain and nervous systems develop the most from 3-7 weeks of age. This is an important stage in their life because social attachments and hierarchal relationships form then. If breeders do not allow their pups to bond with other dogs during this period, they will lack necessary canine social skills later. They will not learn how to accept the roles of submission and dominance.

Photograph 5.1 **Dutch Shepherd**

After 5 weeks of age, pups should be taken from their mothers to begin the human bonding process and the shaping of their behavior through positive reinforcement. However, they should be removed only for short periods of time. Experiments prove that puppies that are completely removed from their mothers too soon to begin the bonding process with people often develop excessively dominant behaviors toward them once they are adults. They may also develop sexual preferences for people. (Pfaffenberger 1963:125). On the other hand, keeping pups from bonding with human beings until they are over 8 weeks old causes them to be wild and independent.

When the dogs are between 8 and 10 weeks old, breeders should expose them to a variety of different stimuli, including livestock, noise and people. They should have taken their puppies for walks through barnyards and

Photograph 5.2 **Schapendoes**

pastures to acquaint them with the sight and smell of livestock. Be certain that breeder has exposed his puppies only to gentle livestock that did not harm them. Otherwise, their herding instinct might have been obliterated.

At 9 weeks of age dogs go through a fear period. If dogs come into contact with anything that causes them fear, that fear becomes an established part of their temperament. This has advantages and disadvantages. If the dog is frightened by livestock, a fear of livestock will remain with it for the rest of its life. If the dog learns not to be underfoot because it fears being trampled, this characteristic may save its life (Lithgrow 1987:194).

Between 12 and 16 weeks, puppies become independent and they begin to get into trouble. This is also when they decide whether you or they are in control. Consequently, training should begin before this critical period ends; you should not wait until they are mature enough to work stock.

John Holmes likens a puppy's development during this period to growing a plant:

"The submissive instinct develops much earlier than the herding instinct. Unlike the herding instinct, which left to develop on its own, will get stronger and stronger, this instinct tends to weaken as the dog grows up. An instinct can be compared to a plant; with suitable treatment it will probably flourish and prove useful. Rough treatment in the seedling stage can easily kill it or so check it that careful cultivation is necessary if it is to be kept alive. The normal herding instinct usually goes through a seedling stage when it can be easily killed or badly damaged by trampling. A little care at this stage will result in a strong, healthy plant which can be trained

Photograph 5.3 Old Fashioned German Shepherd

Photograph 5.4 Belgian Malinois

Photograph 5.5 Beauceron (Arco du Chateau Rocher), Photo courtesy of Breeder/Owner, Karla Davis

Photograph 5.6 Briard

Photograph 5.7 Spanish Water Dog

Photograph 5.8 Polish Lowland Sheepdog, Photo courtesy of Volker Klocke

and molded to whatever shape you want. Generally speaking, instincts strengthen with usage, and if never used, tend to die out. The importance of all this lies in the fact that many dogs which will not work have had the herding instinct killed in exactly the same way as the young plant that is kicked out of the ground" (Holmes 29-31).

After deciding on a breeder, it is time to pick a puppy. First, see if the litter you are choosing from has been temperament tested. Analyzing the test results can narrow your choice. Be aware that such tests are by no means infallible. Puppy tests are done when litters are 7-weeks old. Many breeders use a test developed by Joachim and Wendy Volhard, generally known as the Volhard Puppy Aptitude Test. It evaluates puppies for sociability, dominance and obedience.

To obtain optimal readings, this test should be administered at 7 weeks of age. Studies show that puppies' neurological development reaches adult levels at this age. Consequently, pups tested earlier are immature and pups tested later are often affected by the fear stage that begins at this age. Puppies should be tested by a stranger in an unfamiliar setting. The first five sections of the test evaluate temperament, i.e., how the pup accepts human leadership. The last five sections evaluate obedience and working ability. Be aware the test results do not always predict what traits the puppy will display with another adult. It is wise to check whether dogs from past litters of the same sire and dame did, in fact, mature as the test predicted.

The following is a detailed description of the aptitude test. In the first section the vital parts for herding dogs are the restraint and elevation dominance tests. A pup should resist being held on its back and should struggle somewhat against being raised. A pup that will not struggle may back away from livestock when challenged and a pup that struggles too much may not accept its handler's commands once grown.

In the last section of the test, the most important parts are the sound and sight sensitivity tests. Herding dogs should not be distracted by sound. If a car backfires as you are working your dog or it hears thunder off in the distance it should not stop, leaving you to gather the livestock by yourself. The sight sensitivity test is good for evaluating a dog's sustained interest in livestock. Many dogs are interested in livestock when they move, but lose interest when they stop. This negatively affects their ability to control stock. A puppy showing continued interest in an object once it stops moving probably will make a good herding dog.

Appendix V includes a sample test score sheet. As you look at the sheet you will see a category for scoring. In each test, a puppy will respond with one of the 5 or 6 reactions listed. For example, if your pup scored a "5" during the retrieving test, it means it started to chase the object, but lost interest before reaching the object.

Social Attraction and Following Test

This test evaluates puppies' attachment to people. Before the test begins, the tester marks a starting point on the floor. He then positions himself four feet away and kneels on the floor. Once in position, he asks the pup's owner to place it at the starting point, making sure the pup is facing the tester. He then calls to the pup and encourages it to come toward him. Once the pup reaches him, he praises and pets it. The second half of the test evaluates the puppy's willingness to follow its handler. The tester stands next to the pup, making sure it is watching him. He then encourages the pup to follow, using his voice and/or patting his leg as he walks away.

Restraint and Social Dominance Test

The tester begins the restraint portion of the test by kneeling on the floor. He gently rolls the puppy onto its back and places one hand on its chest, holding it just firmly enough to keep it from escaping. As he holds the puppy in position, he makes direct eye contact, making sure to keep his expression neutral. He holds the pup in this position for 30 seconds. Next, the pup is tested for its willingness to forgive. This is important because puppies that hold a grudge can be difficult to train. Again, from a kneeling position, the tester places the puppy at a 45° angle in front of him. He lowers his head enough for the puppy to lick. Then he strokes the pup from head to tail.

Elevation Dominance Test

This test evaluates a pup's ability to respond to a situation over which it has no control. As the pup is facing away from the tester, he gently picks it up, lacing his fingers under its rib cage. He then raises the pup until all four paws are off the ground and holds it in this position for 30 seconds.

Retrieving Test

This test evaluates a pup's willingness to work for its handler. The tester crumples a piece of paper into the shape and size of a racket ball. Then he kneels on the floor and places the pup between his knees, facing away from him. He teases the pup with the ball and when the pup is interested in it, the tester throws the ball approximately 2-4 feet in front of him. He then releases the pup and as it moves toward the ball, the tester moves back about two feet. If the puppy picks up the ball, the tester encourages it to return to him and praises it if it does.

Touch Sensitivity Test

The puppy's pain threshold is measured by this test. It provides the perspective owner with a rough idea of how easy it will be to control the pup physically through negative reinforcement and/or punishment. Since the development of this test in the 1970s, training techniques have changed dramatically. Some trainers no longer rely on punishment techniques, although negative reinforcement is becoming more accepted. The information acquired from this test is still valuable. When a dog's prey drive becomes overwhelming, you must shock it back into the pack drive. For example, if a dog has a death grip on another animal, you must do what is necessary to make the dog release its grip.

To begin, the tester places the pup on the floor next to him. With one hand, he cradles one of the pup's front paws and with the other hand pinches the pup's paw between its toes. He counts to ten, slowly increasing the pressure. When the pup reacts strongly to the pain, i.e., by biting or yelping, he stops.

Sound Sensitivity

The pup is evaluated for its reaction to loud noises, most often using a metal pan or feed dish and a spoon. When the pup is facing away from the tester, he bangs the spoon on the pan, making a loud, sharp noise. If the pup shows no reaction at all, then it may be deaf. Most puppies, however, will be startled, but they recover quickly.

Sight Sensitivity

A puppy's response to movement is gauged through this test. The tester ties a cord to a towel and drags it past the pup. A pup that retains interest in the towel after it has stopped moving probably will make a better herding dog than one that does not. After all, you want a dog that is always interested in the stock, whether or not they are moving.

Interpreting Test Results

The test is scored using a six-point scale. A pup scoring mostly "1s" is not likely to accept a person as its leader. It is not even a good candidate for a pet.

A puppy scoring mostly "2s" is a self-confident dog that, if provoked, may bite. Experienced handlers can train dogs like this because they will accept human leadership. When working with this type of dog, you must be consistent. If you are an experienced handler, this dog is for you.

A puppy scoring mostly "3s" is outgoing, active, full of energy and always wanting to run and play. As long as they are controlled, they make very good working dogs. If you like training, then this dog is for you. You must start obedience work early, however.

A score of mostly "4s" makes an excellent pet. It is very willing and so easily trained, that you might think it trains itself. This is the dog all pet owners dream of.

When a pup scores mostly "5s", it probably lacks self confidence and will not adjust well to change. It may also be shy and go through life afraid of most things with which it comes in contact. It is best placed in a quiet home.

Mostly "6s" indicates the puppy is very independent and not very affectionate. They know what they want to do and generally resist human guidance in the process. A dog like this needs a reason to obey. It must think an activity or behavior is its idea and not yours (Gazette 1985:42).

Puppy Herding Instinct Test

Although the Volhard test provides helpful information, its results are not inerrant. Therefore, I prefer to administer an additional test when puppies are between 3 and 5 months old. The test works as follows: I set up two adjoining pens. The first, approximately 30-feet-long and between 8 and 16-feet-wide, is the test area, while the second, 8 x 8 feet, holds the puppies. The adjoining, or common side of the two pens, must not form a solid wall because the puppies must be able to see through the divider into the test pen.

I use 30 to 40 pound lambs for the test. I want the puppies to be able to overpower the stock, but I also want the stock to be strong enough to endure the test.

Next, I place the pup or puppies in the smaller pen and bring a lamb or two into the test pen. I carefully watch the puppies to see how intently they watch the lambs as they are brought in. Then I note how long each puppy maintains its interest and which one in the group maintains it for the longest time. I place the puppies into the pen with the lambs. If the pups show a high drive or interest in the animals, I bring in only one puppy at a time. However, I will put more than one pup in if the puppies are young (2-3 months).

Once they are in the pen with the lambs, I evaluate the pups' flight or fight response. If a lamb chooses to fight (manifested by it standing its ground via stamping one of its front hooves) I carefully observe the puppies' reaction. I am looking for those puppies that either stay and hold their ground, maintaining direct eye contact with the stock, or that move toward them. At this point, I evaluate the puppies' response when the sheep move away from them. As this occurs, the puppies should pursue. Hopefully, some of them will attempt to grip a lamb as they overtake it. I am looking to see which pup takes hold in the appropriate locations (see "Gripping", Chapter 9), and whether the grip is proper. In other words, does the pup take a firm hold of the lamb to control it or does it try to rip and tear as it grips? When the sheep stop moving, I identify those pups whose interest in the stock continues. I can't emphasize enough that a pup that is interested only when a lamb is moving probably will not make a good herding dog.

Sometimes, a puppy will bark continuously in the presence of stock. That dog is not a good prospect for high-level competition. If a dog barks, it can't hear my commands and it upsets the livestock but most important to realize is that its barking probably masks insecurity.

The last thing I look for is intensity and focus in a puppy's movement, namely, a pup that moves back and forth in a single track. When grown, that dog will hold a boundary much better than one that creates a wide path. As von Stephanitz noted, "This narrow gait is of great importance for the shepherd dog, because the dog that works sheep and has to keep them off the fields must run in a narrow furrow border of the ploughed field lest he injure the crops himself" (1925:38). Moreover, the wider running dog will be more easily distracted as he works and will tire faster. This brings us to structure. In the pages that follow, we look at some of the physical characteristics to look for in choosing a herding dog.

Chapter Five: Good Herding Dogs

Photograph 5.9 A Belgian Tervuren pup displaying prey drive, Photo courtesy of Kevin Udahl

Structure

As Henry Drummond noted in the 19th century, "Heredity and environment are the master influences of the organic world" (quoted in *Webster's Dictionary of Synonyms*: 1955). The key to their power lies in the process of natural selection. Those members of a species poorly equipped to survive in given environments quickly die, while those better equipped live to pass on their genetic inheritance to their progeny. Wolves, for example, kill animals they can catch, the old ones, those with birth defects, sick and uncoordinated ones. The younger, healthier and naturally faster animals that can outrun wolves escape. Consequently, they produce more offspring and their legacy comes to dominate their species' gene pool. This same process affects the wolf, as well. In order for wolves to survive, they must find and kill their prey. To accomplish this, they must be able to cover great distances in the course of a day. Wolves have evolved into efficient trotting animals with a structure designed for endurance, as much as for short bursts of speed.

Wolves also must be able to change direction quickly. After traveling for miles in search of prey, they must chase, attack and kill it. Von Stephanitz described the structural requirements of this quick-turning ability in his book *The German Shepherd Dog in Word and Picture*:

"The wolf has the powerful long stretched build of the running beast of prey, enabling him to maintain a tireless trot. The high shoulder, the result of well developed withers, gives him greater space for the muscles of the far reaching front members: The hind quarter is perfectly adapted for the proper thrust from behind and below, and also, thanks to the slight arch of the back, permits an enduring and space devouring gallop. Favorable angulation of the limbs gives an excellent reach and therefore swift progress. The chest development is, as with all wild dogs, less prominent, which is the case too with our domesticated dogs, though bred to have a deep chest. The body gives a barrel-like impression, the front limbs appear to the eye to be placed further under the body than is the case with the dog. The front paws, which occasionally must serve the purpose of digging, are, like the front limbs, from the ankle downward, large and extraordinarily powerful. If we examine the track of the wolf, it is seen that he steps with a closer splay of the middle toes than the dog, whose splay has become softer and more yielding. Shepherd dogs, accustomed to strenuous running, and having in consequence well closed paws step considerably more like the wolf…" (von Stephanitz 1925:35).

The point is simple: in nature, function largely determined canine structure. With domestication, however, the process became more complex. Human selectivity became a critical element in the canine evolutionary equation. With respect to herding dogs, the natural environment continued to play an important role in development.

The Louisiana Hog Dog, also known as the Catahoula Leopard Dog, provides a case in point. Getting feral hogs out of the Louisiana bayous was difficult because dogs could not drive them forward like sheep or cattle. Over the years, hog men devised solutions to the problem, all requiring somewhat unorthodox herding techniques. One of these is unique and exemplifies in fascinating detail the combination of heredity, environment and human selectivity in action.

When a Hog Dog came upon a herd of hogs, it challenged them to a fight. At the last moment before the hogs charged, the dog turned tail and ran toward the pen. If a dog misjudged the speed of the approaching hogs or could not run fast enough to escape, he was killed or seriously injured. After about 300-400 yards, the hogs would give up. The dog returned to confront them again. This battle repeated itself until the hogs finally charged after the dog into the pen. To keep them in, however, the gate was immediately closed, leaving the dog in the pen with the hogs. The dog would have been killed, but for a hole or window about three-and-a-half feet off the ground through which the dog could jump.

If a dog was not agile enough to jump quickly through the window, the hogs killed it (Lyon 1973:13-17). The nature of their work, as defined by people, became part of the natural environment that selected for "correct" Hog Dog conformation. McDowell

Lyons relates the following story, which illustrates the process of Hog Dog evolution particularly clearly:

> When we reached Louie's home, a little boy came out to meet us, dragging a huge hound by its ear.
>
> "That's a powerful looking animal."
>
> "Him?" Louie sneered. He's by Old Drive out of Nellie, two of the best dogs ever worked hogs, but that critter couldn't keep out'n a hog's way two minutes. Never would have kept him 'cept the kid raised such a holler."
>
> As the dog turned, his straight shoulders and unbent stifles were apparent. So this swamp man was rejecting the same features that a judge in morning coat should turn down in a show ring. The swamp man knew only by experience that a dog like him "couldn't keep out'n a hog's way two minutes." He did not know the reasons, He had never bothered to analyze them as the bench show judge should do. However, unlike the latter decision, if Louie had not discarded these features, the hogs would have (Lyon 1973:17-18).

Although sheep and cattle dogs did not labor under such hazardous conditions, the environment in which they worked also affected their evolution. Herding dogs were treated as machinery, not as family pets. If a dog could not perform its job because it lacked the right physical capabilities (given its job and the conditions under which it worked) it was replaced by one that could. Sometimes, owners sold such dogs as pets. In most cases, they died a natural death or were culled. Although culling might seem cruel, it mimicked evolutionary processes in the wild. When dogs that cannot perform are allowed to breed, they produce other dogs similar to themselves and the breed, as a whole, becomes less capable of performing the work for which it was originally created.

Because work demands on herding dogs were similar, the dogs that evolved shared certain structural characteristics. However, not all these dogs looked alike. In many instances, differences did not significantly affect the dog's ability to perform its job. In other instances, structural differences were related to the environmental variations in which dogs worked. Even within breeds (the Border Collie provides a good example) different types emerged.

As herding breeds became popular with "pure bred" dog organizations like the American Kennel Club, much of the variation within breeds disappeared, along with working ability. This happened because written conformation standards, rather than working environments, determined which dogs were bred. Too often, these written standards lost sight of how function and form should interrelate. Even when they did not, written standards were not as effective as ruthless environmental demands in producing good working dogs.

This is because conformation standards, no matter how well written, allow various interpretations of herding dogs' appearance. Inevitably, a dog's appearance becomes a critical factor in breeding practices. Breeding that is designed to conform to written standards produces dogs with the right herding "look." Unfortunately, the ability to perform useful work became less important. The movie "Babe" provides an excellent example. The dogs used in close-up shots were "show" Border Collies, dogs bred to have the correct herding "look." These dogs couldn't do much actual herding and real, working Border Collies had to be used in scenes that involved work. Because the working dogs didn't have the right "Hollywood look," they weren't used in close up shots.

Even though structural faults are listed as such in written standards, sometimes, little weight is placed on their importance. As Lyon points out, a dog can have a major structural fault" and win in spite of it because the others are just as bad or because the fancy at the time is emphasizing special characteristics. In such a case the breeder is apt to form a mental blueprint that is not sound, and as time passes the fault becomes an earmark of the breed or of his own animals" (Lyon 1973:4).

Written standards often place undue importance on traits that are unrelated to working ability. For example, the Australian Shepherd standard faults prick ears because they allegedly lack "protective hair that prevents burrs and other debris from lodging in the ear."(Hartnagle 1985:12). Yet, prick-eared herding dogs have demonstrated time and again that their ear set is unrelated to their performance in working environments. Focusing on ear sets can lead breeders to ignore more critical herding characteristics. As a consequence, the herding abilities of their dogs degenerate.

Photograph 5.10 "Magic" a Winston Cap type Border Collie.

Photograph 5.11 "Fly" Nap type of Border Collie

You can still find breeders of working dogs, (usually people who are primarily farmers, ranchers or shepherds themselves), for whom the true test of "correct" conformation is a dog's ability to do the required work. Their breeding programs mimic natural selection, i.e., the working environment "selects" the dogs they choose as breeding stock. They only breed dogs that prove they have the necessary conformation and instinct to accomplish the job at hand and to do it over an extended period of time. Sometimes they have to work at least two years before qualifying as breeding stock. For these breeders, function drives form.

Although choosing a puppy is always a gamble, such breeders, offer the best chance of getting a dog with good structural characteristics. Unfortunately, like good dogs, these breeders are often hard to find. In any case, buyers wanting the best dog should be aware of the structural characteristics that most relate to a dog's ability to do the work required of it. Accordingly, the pages that follow discuss in detail these structural characteristics.

In the course of a tending dog's day he must perform a variety of tasks. First, he must move efficiently over an extended period of time without tiring. Second, he must change direction quickly to control his charges. Finally, he must jump into the sheep pen. The next section describes the elements of a tending dog's structure that are important for these activities. Let's begin with movement.

Gaits

Dogs move in a variety of gaits including walk, pace, trot, and gallop to name a few. Unlike either the walk or gallop, however, both the trot and pace use energy economically and both allow reasonable speed over long distances (Gilbert and Brown 1995:20). The trot, however, is the more efficient gait for tending dogs, given the nature of their work. The Briard standard even states: "He is above all a trotter ..." (AKC 1979:308).

The trot is comprised of two diagonals, left and right. When a dog trots, his left front and his right rear form one diagonal while the right front and the left rear form the other. When one diagonal is on the ground, the other is in the air. Nevertheless, the front foot is raised off the ground just before the rear.

Balance plays a key role in the way in which dogs move. One of the laws of motion is that in order for a body to move forward efficiently it must do so in a straight line. For example, when a dog is stationary his legs are spread apart allowing him to stand still, but when he begins to move, this stance is no longer efficient. He must now move his legs toward the center line of his body. Consequently, the tracks left behind are in a straight line.

Tracking or Lateral Stability

As dogs move, their bodies begin to move side to side and in some dogs this is more exaggerated than in others. Dogs that move with more sway use greater energy than dogs that have narrower, single tracking movements. This sway reduces a dog's endurance because it causes him to use energy while compensating for lateral instability.

For example, when a woman is in the last stages of pregnancy she must walk with her legs spread farther a part than she does in the earlier stages and because she is walking with her legs spread apart she tires faster. This is because she exerts more energy as she rolls from left to right trying to maintain balance.

Single tracking, however, allows an animal to move forward in a straight line without side to side movement. This is best observed in the wolf: A wolf's hind feet step into the tracks made by his front feet. In contrast, a dog's hind feet leave impressions next to his front feet; the hind leg crosses over this center line to the side of the opposite front leg (Von Stephanitz 1925:35, Mech 1970: 13).

Although even dogs with excellent conformation do not single track, they do begin to approximate single tracking at increased speeds and of course, those who approximate it more closely expend less energy than those who do not (Lyon 1973:90). Therefore, when evaluating herd dogs for their ability to approximate single tracking, they are best observed at a fast trot.

A dog's tracking ability is best evaluated when it is mature enough to outrun livestock. I find it easier to identify a straight tracking movement if the dog is moving near something that is straight like a fence or a natural demarcation. Watch your dog run along a fence line while the stock is on the other side or along the edge of your garden. Then look for evidence of a single tracking movement when you start working your dog along a boundary line during training sessions. As your dog moves, his path back and forth should be narrow compared to a wide path that needs more space.

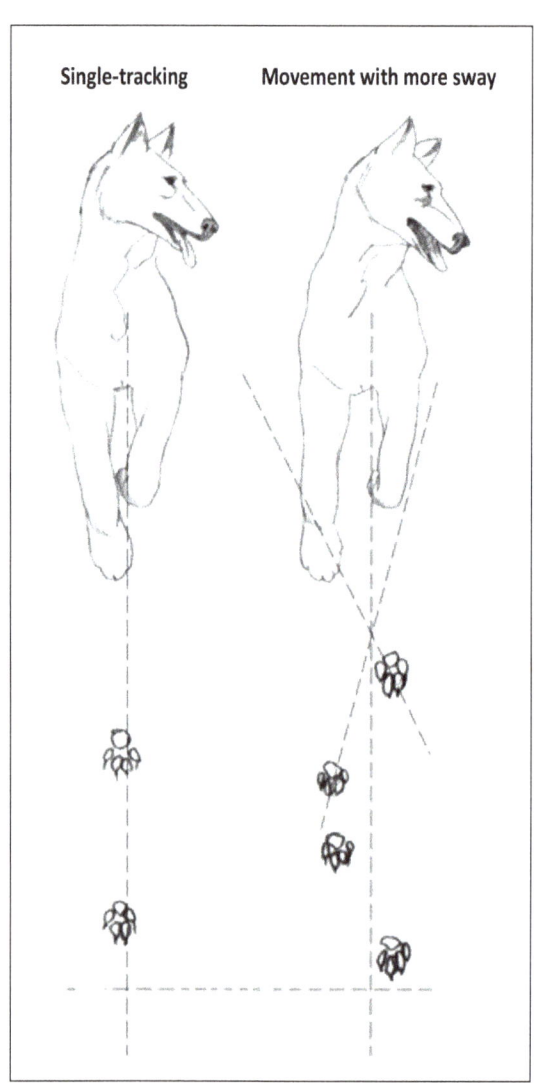

Figure 5.1 **Tracking**

Turning

Like the wolf, herding dogs need to turn and change direction quickly. As the Briard standard states: "His movement has been described as "quicksilver," permitting him to make abrupt turns, springing starts and sudden stops required of the sheep herding dog" (AKC 1979:308). In order for a dog to turn quickly, though, he must turn on his hind quarters rather than his front end. Many of us have seen Quarter Horses either cutting cattle or competing in reining competitions. Both events require horses to turn and change direction very quickly. To do this they must shift their weight to their hind end, causing them to dig their rear legs into the ground. This action frees the front end, allowing it to rise off the ground. The horse then rolls back over his rear legs, spinning in a different direction. In contrast, a dog or horse that turns on his front end, walking his rear end around, cannot turn as quickly.

Jumping

Although jumping into a sheep pen may only be a small portion of what tending dogs must do in the course of a day, it is very important that their structure allow them to lift their bodies off the ground, to horizontally propel them over the fence, and to absorb the shock of landing. Consider the sequence of movements necessary to accomplish this:

> As the dog approaches the jump, he places his front feet, one slightly ahead of the other, at a take off spot on the ground in front of the jump. Then, as the front legs are planted, the dog lowers his head and flexes the front legs a little. He then flexes his spine and brings the rear legs forward, placing them on the ground slightly ahead of the front feet. The dog then extends the front legs, pushing the front of his body upwards and raising his head to assist with upward thrust. The rear legs are then extended, propelling the dog's body upward and forward over the jump. Once the dog is in the air, he lowers his head closer to the outstretched front legs in order to help with forward thrust and to reduce drag. At the apex of the arc, the dog should lower his head and lift his tail to help rotate the body forward and down. He stretches the front legs forward and down, one ahead of the other, to prepare for the landing. After the front legs have hit the ground, the rear legs are drawn forward under the dog's body to absorb some of the impact of landing (Zink and Daniels 1996:21-22).

In the pages below we look in detail at the structural characteristics necessary for dogs to efficiently make these movements.

Static Conformation

Whether a dog is trotting, turning, or jumping, his form must follow his function. In other words, static conformation affects a dog's ability both to maintain balance and move efficiently. Thus, it is important to take a closer look at a dog's static conformation and its effect on movement.

Forequarters

The front quarters of a dog are comprised of the shoulders, upper arms, forearms, wrists, pasterns, and front feet. According to canine structural experts, they also carry 60 to 65% of the dog's weight (Elliot 1985:41; Zink and Daniels 1996:9). The shoulders have several functions. They support the dog's weight, work as shock absorbers, propel motion through turns, offset lateral displacement and help maintain the dog's center of gravity (Lyon 1973:130).

The blade of the shoulder must rotate 15° in either direction to allow the front legs to move back and forth. A trotting dog should have a shoulder lay-back of 50-60° (Gilbert and Brown 1995:64). To measure this place one finger on the top of the shoulder blade and another at the point of the shoulder. Draw an imaginary line joining the two points. Then draw an imaginary line from the point of the shoulder to the elbow. Both measurements should be equal (Smythe 1970:68). Moreover, the length of the shoulder blade and the upper arm should be the same and the angle between them should be 100-120° (Gilbert & Brown 1995:74).

If the shoulder is too steep the dog's ability to reach is diminished. This also causes dogs to move with short choppy steps and causes their withers to move up and down. Both are faults which weaken endurance. When a dog's shoulders are set too far forward, its shoulder blades point toward each other. This causes its elbows to turn out and its toes to turn in. This can also cause the front feet to move in a circular motion away from the body at the end of each step, a condition called "paddling" (Spira 1982:69). Paddling wastes motion and causes dogs to become fatigued.

When dogs have narrow shoulders they have difficulty lowering their heads, which in turn affects their ability to turn and jump. You can test whether a dog has shoulders that are too narrow by placing the tip of one finger in the middle of the withers and pressing downward. Your finger will pass between the scapula to the dorsal spinous process. Now lower the dog's head toward the ground. If the dog's shoulders are too narrow the tip of your finger will become pinched by the scapula (Smythe 1970:82).

When observed from the front, the forearms should be straight, i.e. dog's elbow

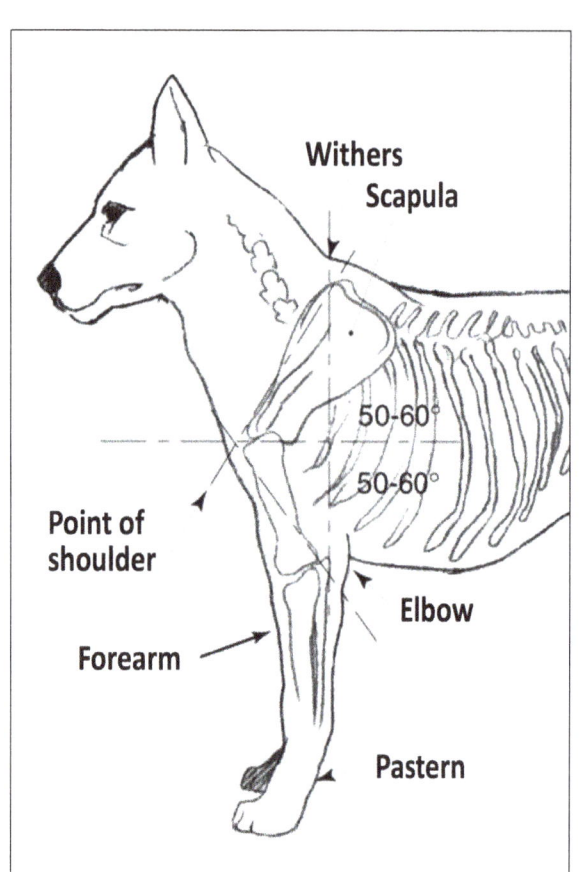

Figure 5.2 Forequarters

should be parallel to its body. If a dog's elbows are not in proper alignment, they will point outward causing his paws to turn in toward his body. This is commonly called being "out at the elbow." If the opposite is true, and the elbows turn in while the paws turn out, the dog has a "fiddle front." Again, these faults diminish a dog's ability to move efficiently.

When viewed from the front, the pasterns should appear straight but from the side, they should have a slight angle. Like the shoulders, pasterns act as shock absorbers. As a dog's front foot hits the ground, the shock of the concussion passes from the pad to the pastern. "Weak pasterns predispose to hyperextension of the pastern and knee (carpal sic.) at fast speeds, thus inducing injury. Unyielding, straight pasterns with no shock-absorbing qualities jar the entire system and lead to early fatigue" (Hartnagle 1985:15). Weak or soft pasterns cause a dog's weight to be carried by his sinews and ligaments (Sárkány and Ócsag 1987:37) rather than by the heel of the pad. This causes a dog's feet to break down (Lyon 1973:147).

Pasterns that are too straight cause "knuckling over." In other words, the pastern joint (carpal) is bent forward instead of forming a vertical line. This keeps pasterns from absorbing the pounding of the body on the ground as the dog moves, which leads to fatigue (Hartnagle 1985:14)

You can experience this yourself in the following way:

Place the knuckles of your hand on the table top and press straight down on them. Hold the wrist joint, which is the same as the dog's pastern, so that it forms a vertical line. In this position it does bear weight without effort. Now move it slightly forward out of line and notice the quiver that comes into it, which is what happens when you see quivering pasterns (Lyon 1973:154).

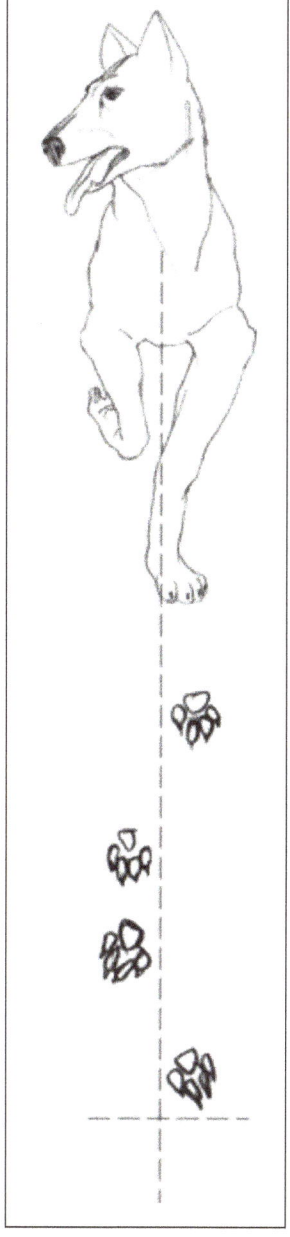

Figure 5.3 Paddling

Many confuse the term "front," with a dog's front quarters. The term "front," however, refers only to a portion of the front quarters, i.e., from the elbows to the feet as viewed from the front. Dogs have many different types of fronts including bowed, narrow, wide, east-west, fiddle, and pigeon-toed. They have bowed fronts when their forearms curve outward from the elbows and bend in at the pasterns; they have narrow fronts when their forearms are close. An inadequate spring of the rib cage is also associated with narrow fronts.

Dogs with narrow fronts have diminished heart and lung capacity which affects their endurance. Dogs with east-west fronts have pasterns that cause their front feet to turn out, as do fiddle fronts. But the fiddle front, unlike the east-west front, is shaped like a fiddle, i.e., the elbows are set wide apart while the forearms bend in, making its shape similar to a fiddle. Finally, dogs with pigeon-toed fronts have pasterns and feet which turn inward.

Each of the above faults causes dogs' fronts to be out of balance with their rears. As a result, they cannot approximate single tracking, which in turn leads to greater fatigue. This is because when the movement of the front and back are out of adjustment to one another, dogs' rear legs can hit their front legs. They can only avoid this by placing their back feet either inside or outside of their front feet. In other words, they must double rather than single track.

If a dog has poor conformation in the front, it will be obvious in the movement of the animal's withers. Withers that move with lots of bounce mean the dog's shoulders are not absorbing the shock of the front legs as they hit the ground. This causes the animal's front end to deteriorate over time and ultimately causes lameness.

We can better understand this process by using the analogy of an automobile's suspension system. Think of a dog's shoulders as the shock absorbers. When they are working properly the ride is smooth, but when they wear out the ride becomes uncomfortable and tiring. Moreover, the longer you drive a car in this condition, the more damage you do to the car's suspension system.

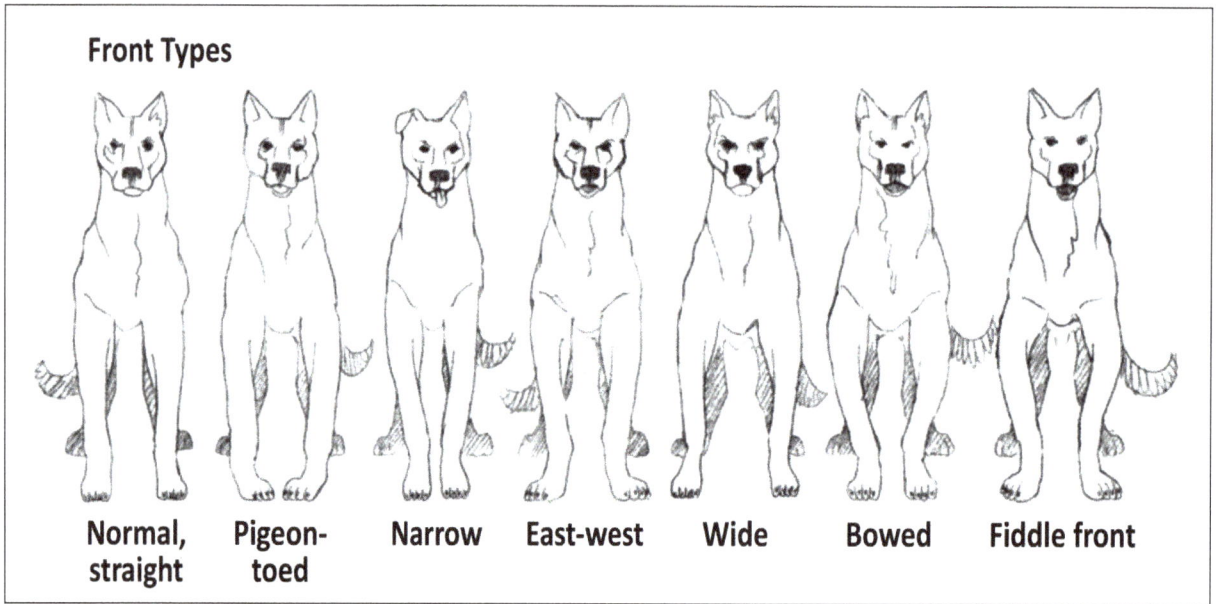

Figure 5.4 **Fronts**

Hindquarters

The function of the hind and front legs are vastly different. The front legs carry the weight of the dog while the rear legs propel the body forward. "You can check this by pressing down on your dog's loin or croup and then on the withers to compare the difference in stability of the two" (Lyon 1973: 189-90). The front will withstand more pressure or weight than the rear before succumbing to the pressure.

The hindquarters allow dogs to turn and change direction. In order for dogs to turn properly they must get both rear legs underneath their bodies. To accomplish this they must possess sufficient forward reach. Alternatively, dogs must have good backward reach which you see in the backward swing of the leg after the foot leaves the ground. Sufficient backward reach is necessary for speed and endurance.

Angulation also relates to speed, endurance and agility. It is determined by the slant of bones and degree of angle in the joints (Elliott 1985:41). A dog's pelvic slope should be 30° off the horizon for proper angulation. A steep pelvis (45°) gives a dog very little backward reach. As a consequence, the dog will move with short choppy steps. Just as important, dogs with steep pelvises must take more steps to make up for the shortness of their stride. This causes fatigue. The length of the pelvis is also important. When it is long it allows more room for the attachment of the rear muscles, thus allowing dogs to jump more efficiently.

The pelvis (which some confuse with the croup) is the section of the hindquarters starting at the loin and ending at the tail. Although the pelvis is a part of the croup, the croup starts above the rear legs at the loin and extends to the point at which the tail begins.

The angulation of the croup determines tail set. A low tail set generally indicates a steep pelvis and thus a short backward reach. McDowell Lyon states, "A steep croup enables a dog to get his feet under him and turn more quickly. The forward reach from his croup draws the weight to the rear hand (hind end), forces the rearing muscles into action, lets the back feet act as a pivot on the turn and directs the force more vertically" (1973: 193-94).

In contrast, a tail that is set high indicates a flat or moderate slope of the croup. A flat croup increases the length of muscles from the croup to the stifle, thus lengthening a dog's back reach. This in turn gives him more power (Lyon 1973:194). When an animal's croup is flat, its rear turning ability is diminished, thus causing the animal to turn on its forehand (front quarters). This diminishes its turning speed. As you can see; each type of croup has advantages and disadvantages.

In recent years breeders have thought that more angulation was needed in the rear quarters, just as a man pushing a car from behind has to extend his arms toward the car and his legs backward. Therefore, breeders began to increase rear angulation by breeding for rear legs that were more behind than beneath their dogs. Theoretically this increased dog's power (Smythe 1970:116) but it was discovered that this lengthened the tibia. In so doing it put added strain on the short, round ligament (teres) that holds the head of the femur in the acetabulum or socket. Thus, as Elliot notes, "excessive angulation in any part of the dog's body is detrimental to joint support and endurance. It is never a question of the more angulation the better—it is a matter of

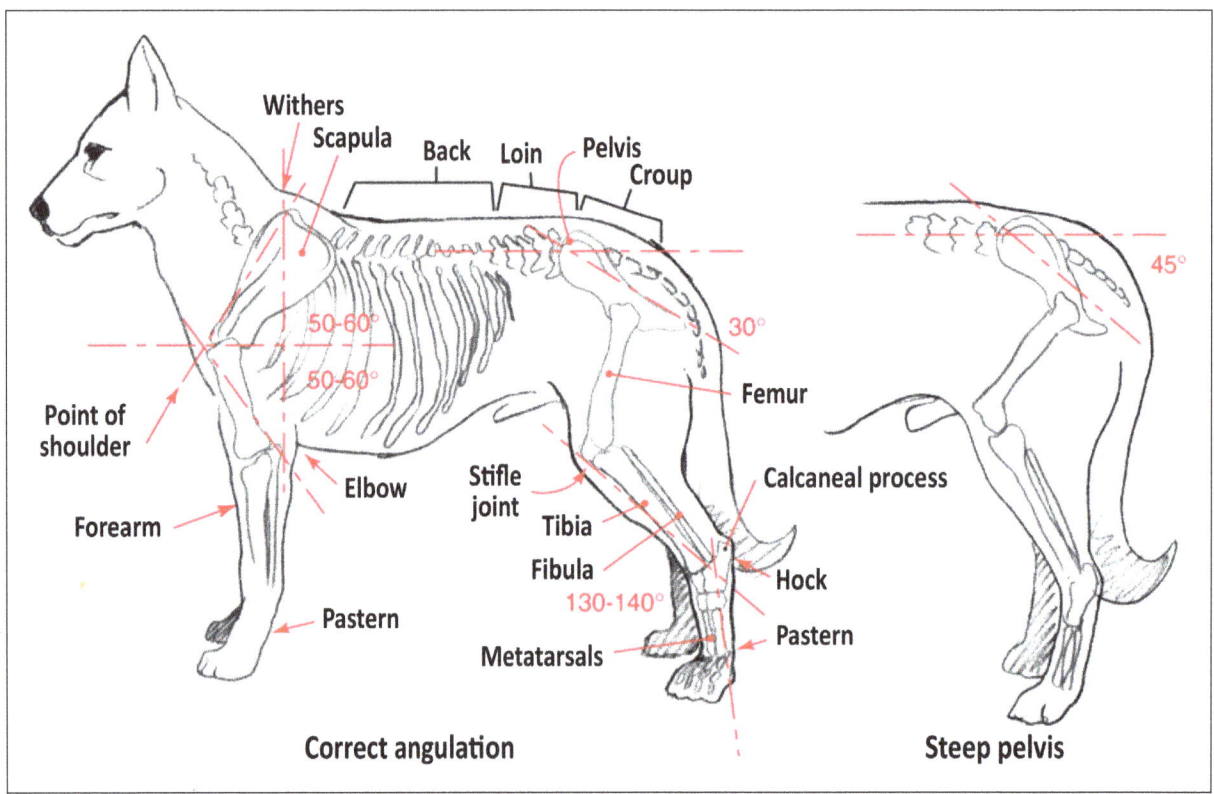

Figure 5.5 **Angulation**

just how much is needed for functional efficiency" (1985:41). Moreover, when a dog has either long rear legs or excessive angulation, he cannot trot as smoothly as a dog that has legs of equal striding length (Gilbert & Brown 1995:23). In spite of these problems many people breed their dogs with long rear legs and excessive angulation because the outline which results is pleasing to their eye.

The angle of the hock joint (the lower joint of the rear leg where the tibia and fibula bones join the tarsal bones) should be between 130° and 140°. The point of the hock is called the calcaneal process. The higher the hock joint is from the ground, the shorter the calcaneal process is. This means that relatively short bones below the hock mean relatively long bones above it. When a dog has high hocks (hocks further from the ground) he has better speed; when he has low hocks he has better endurance. Thus when a dog's job requires him to trot he should have his hocks "well let down" (close to the ground).

Hocks take various shapes—sickle, straight, well bent, cow and high. The term "sickle hocks" comes from the shape of a sickle, a tool with a sharp curved blade attached to a rigid handle. When viewed from the side, sickle hocks are bent more than 90° to the ground with a joint angle of 130-140° (Spiria 1982:80). Sickle hocks cause dogs to move stiffly with short shuffling steps. This reduces their speed. Moreover, they cause dogs to stand under themselves, which gives the impression to the novice that they have well let down hocks.

Chapter Five: Good Herding Dogs

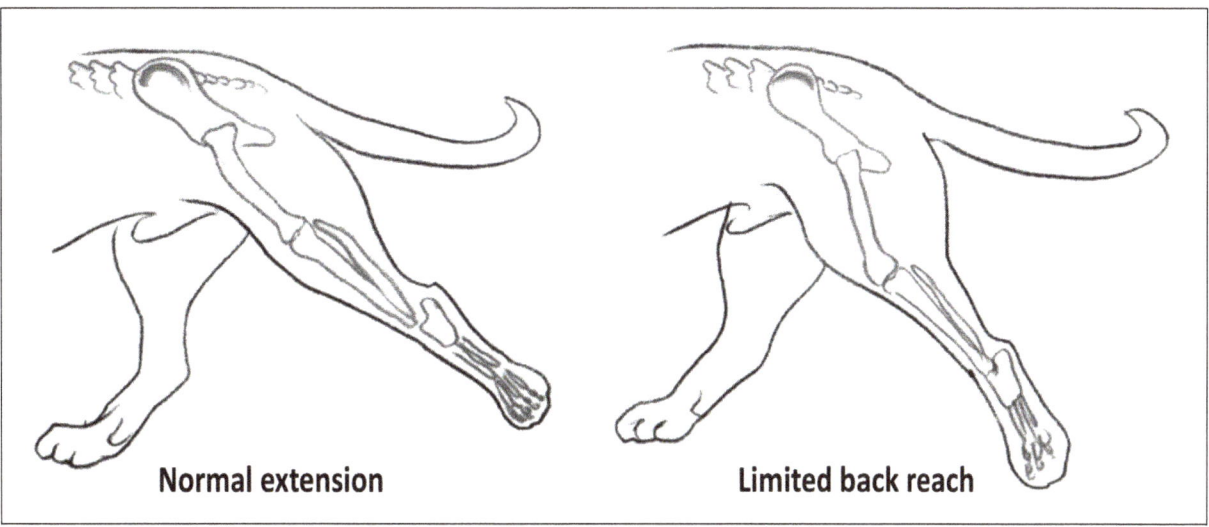

Figure 5.6 **Rear extension**

In contrast, straight hocks have an angle of 140° or more. Cow hocks turn in instead of pointing directly back in line with the body while bow hocks turn out. Both cause dogs' gaits to be out of balance. When a dog's movement is out of balance, i.e., not confined to a single plane, his gait is inefficient. When a dog is high in the hocks, his rear pasterns are long, making the distance between the ground and the hocks

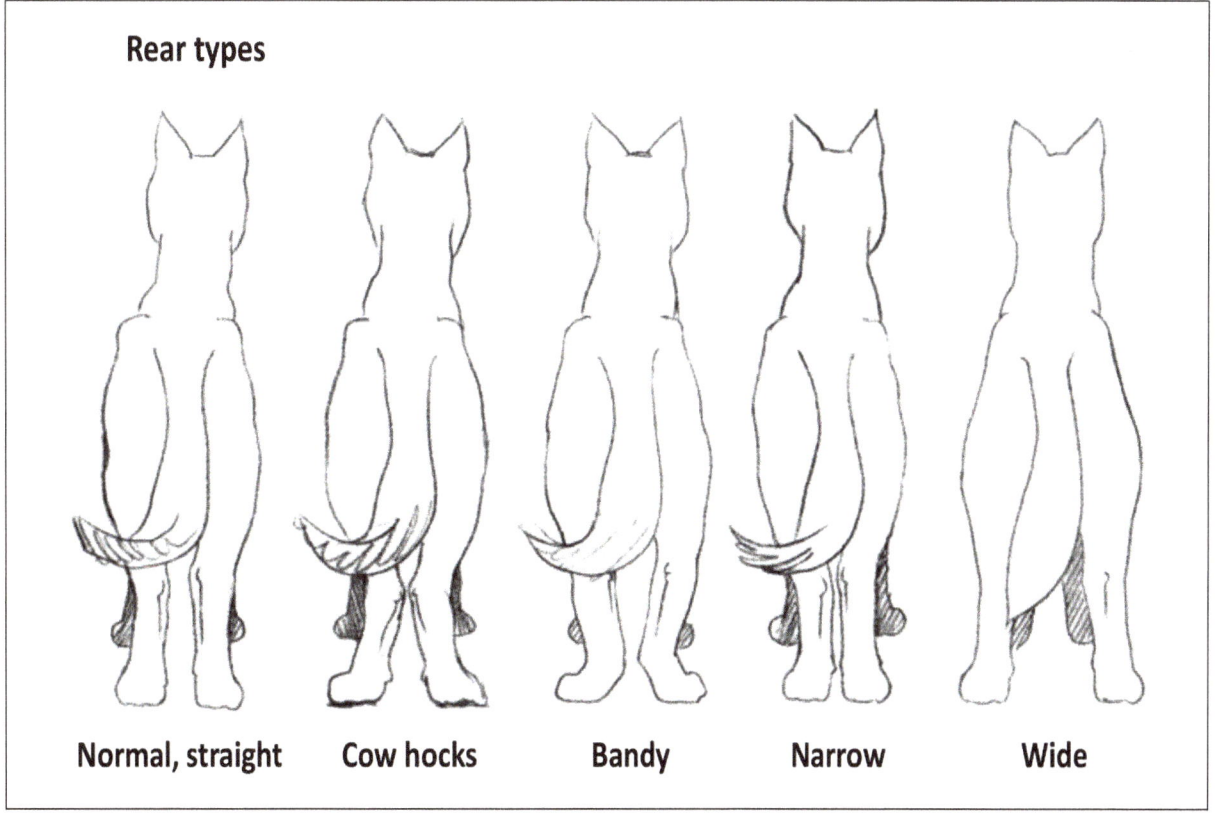

Figure 5.7 **Rears**

99

long as well. This increases jumping power. In contrast, well let down hocks provide endurance. In any case, whether they are high or "well let down," tending dogs' hocks should be well bent or angulated. (This relates to the angle formed between the tibia/fibia and the pastern.)

The stifle joint is located between the upper and lower thigh. When dogs' stifles are well bent they have increased flexibility in their legs. This gives them more reach, which in turn increases their jumping ability and speed. In contrast, dogs with straight stifles cannot efficiently move their rear legs under their bodies, because straight stifles hinder their jumping ability.

Bone

In order for dogs to run fast, the weight of their leg bones must be as light as possible. Oval bones have a distinct advantage over round ones, for "… making the leg bone oval with the long axis toward the front of the dog, the front and rear edges are reinforced without adding extra weight of the bone on the sides" (Gilbert & Brown 1995:78). To tell whether or not a bone is round or oval, run your hand along its front and side. If it is round, both sides will be roughly equal in size. But if it is oval, the bone will be narrower in the front and wider on the side.

Neck

Dogs use their necks and heads to shift their center of gravity. This helps to maintain their balance. For example, when dogs run, they extend their head and neck forward. This in turn, puts more weight on the front quarter, which increases speed. When dogs want to stop, they throw their heads upward, thus moving their center of gravity to the hindquarters, which can then act as a break. Therefore, it is important for the neck to be strong enough to withstand the pull of the shoulder muscles. Strong necks appear arched or convex, breaking at the pole (the juncture of the second and third cervical vertebrae). In contrast, a weak neck (called a ewe neck) is concave. When dogs have ewe necks the ligaments that control the break of the pole are weak and cannot withstand the pull of the shoulder muscles.

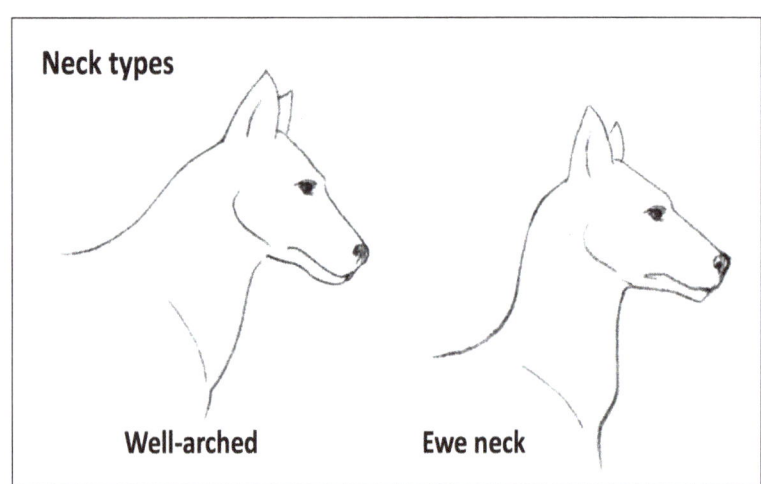

Figure 5.8 Necks

Back

Dogs' backs should be straight, level and firm, not sagging or weak. A weak back lacks muscle and will break down over time. But a straight back does not mean that the spine is level; it means the muscles surrounding the spine have a straight and level appearance. A curved back line causes a loss both in endurance and speed. Although a roached (convex) back is less favorable than a straight back, it is still stronger than a sagging back. A short back makes locomotion hard and uneven, diminishing their ability to turn (Von Stephanitz 1925:535).

For canines that primarily need to trot, i.e., wolves and tending dogs, long bodies are advantageous. Long bodies allow the rear feet to have greater reach without the front and rear feet colliding and thus precluding single tracking.

When dogs have short backs and equal leg length (in both the front and rear) their rear legs also have a tendency to hit their front legs. To avoid this, the dog must move like a crab and as a dog crabs his spine is turned sideways and does not point toward the direction of travel.

Loin

Most standards call for slightly arched loins, not to be confused with roached backs which do not include the loin. The most important function of the loin is to provide greater flexibility in the back. When the loin is flexible it allows a dog to get his rear legs well under his body when turning, jumping and galloping. The loin provides structural strength to a dog's back.

Tail

Many herding breeds do not have tails: the Australian Shepherd, Corgi, and Old English Sheep Dog, to name just a few. Obviously, for these breeds the tail cannot affect movement. Among other breeds the tail functions in conjunction with the head and neck to maintain balance (Gilbert and Brown 1995:136). Tending dogs' tails serve as rudders, especially in making sharp turns (Von Stephanitz 1925:530).

The tail is also related to the strength of the muscles attached to the pelvis. For example, a squirrel's tail indicates that the muscles along the base of the tail have little tension, which in turn, lessens the arch of the loin and thus, the flexibility of the back. When a dog carries its tail high, however, these same muscles are tense and the loins arched. If its tail is twisted, it indicates that one of the two muscles on the top side of the tail is not functioning properly (Lyon 1973:225).

Each of these tail sets causes herding dogs to loose their balance when making sharp turns. They also affect sheep's response to dogs. For example, when tails are held high over the back (gay) it gives the impression that dogs are only playing.

Rib Cage

Endurance depends on the amount of energy expended. If a dog's body does not have the room to properly house the heart, lungs and abdomen, these organs cannot function effectively. Therefore, the bones and tissue that frame this area are as important to endurance as are the organs themselves (Lyon 1973:228-9).

A dog's heart lies on its sternum between the third and eighth ribs. At the eighth rib it tips downward and thus, needs the most room. Therefore, the rib cage must begin its upward swing or tuck after the ninth rib. It is here that the diaphragm separates the abdominal from the thoracic cavity. The length of the diaphragm (which runs from the loin to the seventh rib) is important. Long diaphragm muscles provide the heart and lungs with more space than shorter muscles. This allows these organs to easily expand, which in turn, allows animals to work longer without changing their pace (Lyon 1973:228-231). Therefore, the depth of the rib cage should be measured vertically from the ninth rib to the spine. It should not be measured behind the elbows.

When you measure the depth of the rib cage behind the elbows, the reading is misleading and this may fool you into thinking that the heart and lungs have sufficient space, when in fact they do not. When the ribs begin to shorten at the third or forth rib, two endurance-reducing defects occur: the heart and lungs are robbed of space, and the diaphragm is shortened, making breathing more difficult. Both weaken a dog's performance. This fault is commonly called a herring gut. The novice can confuse a herring gut with proper bottom line tuck-up which refers to the normal abdominal tuck after the rib cage.

All ribs are attached to the spine by ball and socket joints. As dogs inhale, the muscles cause the ribs to rotate, thus allowing the chest cavity to expand. True ribs (1-9) have less effect on chest cavity size than false ribs because they are attached to the sternum, making them more rigid. The false (10-12) and floating ribs (13) are less rigid than the true ribs because they are either attached by cartilage to each other or are completely unattached. Thus, it is the rearward ribs that can expand the most. By doing so, they allow the chest cavity to increase in size, which increases a dog's ability to breathe.

The layback angle of these ribs is important. When the angle is roughly 45°, it allows the longest possible rib to fit into the thoracic cavity without changing the dog's body size. If, however, the ribs are set at a 90°, their rotation does not increase lung capacity (Gilbert and Brown 1995:116).

When ribs are round they form a barrel rib cage. Although this type of rib cage may be desired in a bulldog, a rib cage comprised of flat ribs results in a deep, narrow body which is better for herding dogs. This is because it provides more room for the heart and lungs and aids in combating lateral displacement in locomotion (Lyon 1973:237).

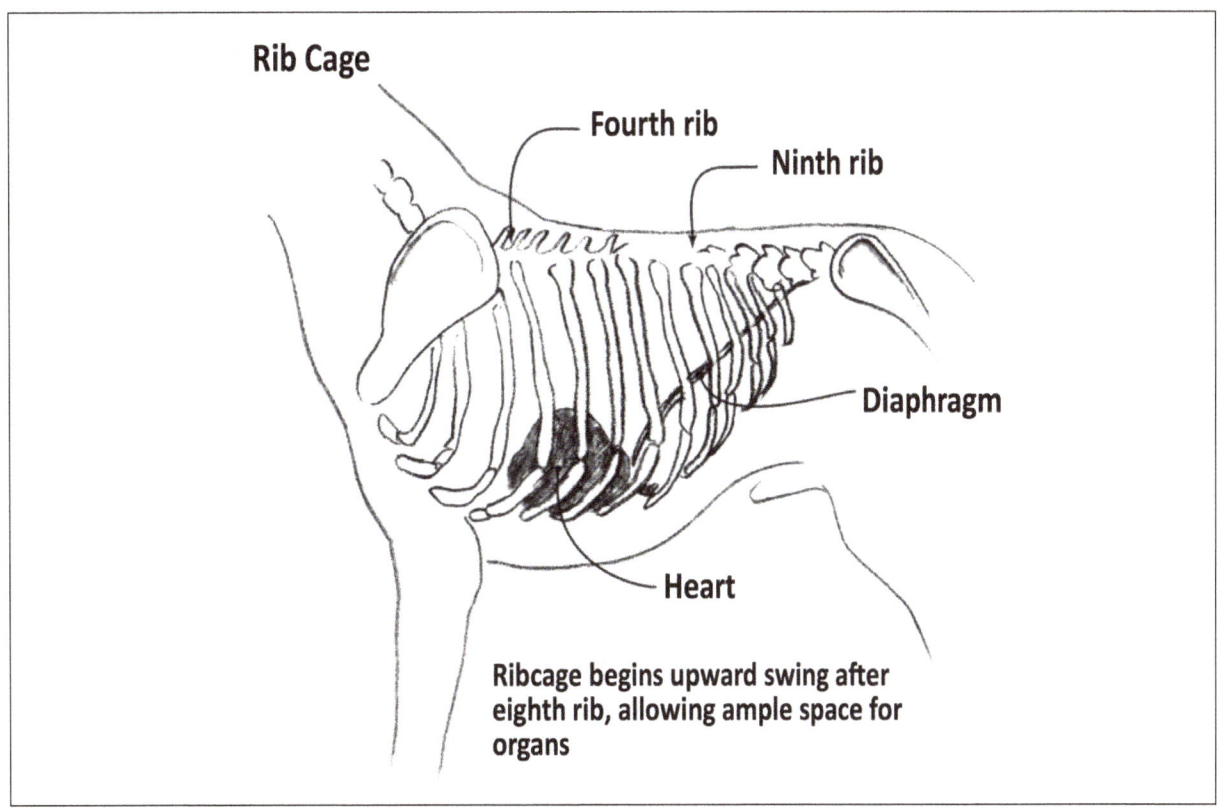

Figure 5.9 Ribcage

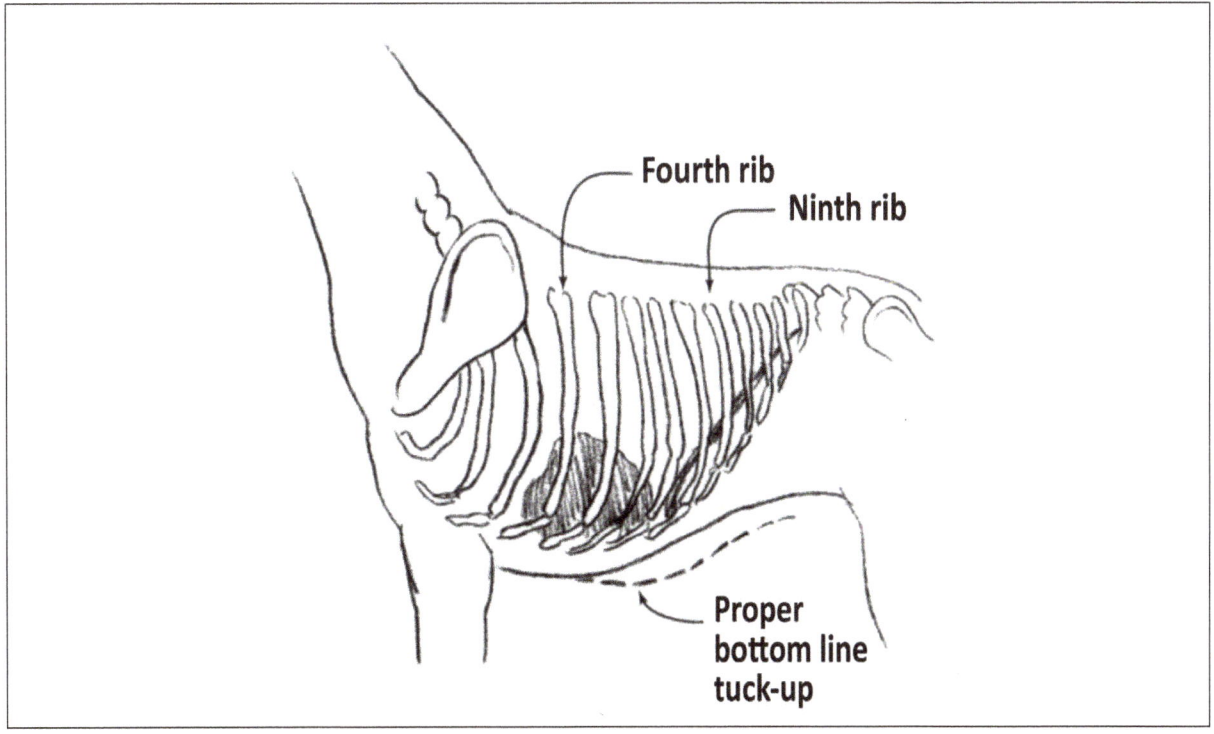

Figure 5.10 Herring gut: rib cage begins upward swing at sixth or seventh rib. This limits the space for heart, lungs, and diaphragm.

Feet

A herding dog's feet should be strong, round, tight, compact, and well-arched and have thick pads. "Weak" feet (i.e. feet with thin pads) are more easily affected by rough terrain, rocky surfaces, briars, thorns, etc." (Hartnagle 1985:15). In such conditions, weak feet quickly break down and the pads begin to bleed. Dogs with such feet need added protection (bandages or leather boots) if they are to work effectively.

The shape of the feet, as well as the thickness of the pads is important. "Cat" feet are round and compact with the two center toes slightly longer than the rest. This type of foot is excellent for endurance work. "Hare" feet have center toes much longer than the rest, making all the toes less arched. "Hare" feet increase leverage and consequently increase speed, but they also increase fatigue (Lyon 1973:156). Consequently, this type of foot is not suited for dogs that must travel long distances while tending their flocks.

Feet that are spread apart and lack compactness are called "splay" feet. They should not be confused with webbed feet typical of water retrieving and sled dogs. Toes set wide apart and often showing daylight between them typify "splay" feet. "Splay" feet disadvantages herding dogs because, like weak feet, rough terrain easily injures them. Stones and other objects often work in between the pads causing soft tissue injuries.

Teeth and Jaw

Herding dogs need strong, properly aligned teeth to control livestock. Properly aligned teeth on a herding dog form a scissors bite,[25] i.e., the lower incisors are in back of and against the upper incisors. This makes the dog's grip more secure. Moreover, a scissors bite minimizes wear and tear on teeth. Whether or not a dog will have a scissors bite as an adult is sometimes difficult to determine when it is a puppy because its teeth are temporary. As a puppy grows the spaces between its teeth become wider, and dogs that appear to have a scissors bite as puppies may not as adults. Therefore, it is best to look carefully at the teeth of the parents before buying a puppy.

Herding dogs should have even jaw alignments. Their jaws should be neither overshot nor undershot. When a jaw is undershot, the lower jaw protrudes forward, creating a space between the upper and lower incisors. When the jaw is overshot, the

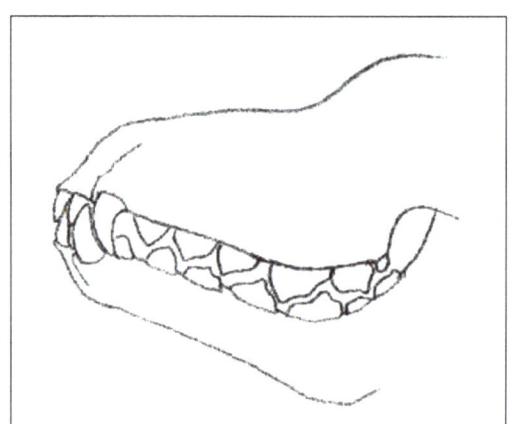

Figure 5.11 **Proper scissors bite.**

---

25 Upon "closing the blades of a pair of scissors the blades slide past each other, super-imposing one behind the other" (McLeroth 1982:97)

upper jaw protrudes forward with the same result. In both cases, dogs cannot properly grip livestock.

In looking back at canine history many enthusiasts conclude that the dogs of today are better than those of the past but this is not necessarily true, especially with herding dogs. Times have changed greatly in the last 150 years and most dogs are no longer used in the ways for which they were originally bred. "This change has taken our dogs out of the fields and off the road and put them in backyards and living rooms, leaving the dictates of their characteristics almost wholly in the hands of fancy" (Lyon 1973:248). As standards have changed, many dogs have become unable to perform herding maneuvers as well as their ancestors did. Over the years, individual dogs of some breeds have become larger and cannot dodge a cow's hooves as quickly and easily as the smaller ones of its breed. The Austalian Cattle Dog is a case in point. Although many breeders and judges today do not consider this a problem, it can be for working dogs.

I have observed several of these large ACD's working cattle and each time they saw a cow lift its leg to kick they instinctively tried to duck under its hoof. Unfortunately, they were continually kicked in the head because they were too big and too slow to get out of the way. These dogs were handicapped because their instincts told them to work in ways their bodies could not. If, however, they were smaller in size, or their instincts were like the large cattle herding breeds (e.g. the Bouvier) they would not have been so easily kicked.

Years ago farmers and ranchers bred their dogs exclusively to make their work easier and as long as their dogs could do their jobs effectively they survived but as the breeders changed their breeding practices to achieve a different, standardized type, structural faults sometimes emerged in the process. These structural faults, at first only tolerated, often became the norm later. Today, finding dogs with good structure is often difficult. The information provided here, however, will help you choose a dog that can effectively execute his instincts. Once you have a good understanding of breed and herding type, structure, etc., you can start your search for a puppy. After such careful selection, you must consider the health and socialization of your pup very seriously.

Caring For Your Puppy At Home

Once you have chosen your puppy and brought it home be sure to have its health evaluated by a veterinarian. If it is unhealthy, return the pup to the breeder for a replacement or refund. Because sheep are range animals and may come in contact with rabid wildlife, your pup must be vaccinated against rabies before it is introduced to livestock. (Be sure to vaccinate your sheep against rabies as well.) When the pup is older it should have its hips and elbows x-rayed for dysplasia since many herding breeds have a genetic proclivity for this disorder.

If you have selected two pups, be sure to separate them. Keeping them together in the same pen can adversely affect their development. By separating them, each can develop its own personality.

One last thought. Your goal is to choose a dog with good structural characteristics but "good" is always a relative term. It is just as important, if not more so, to choose a dog easy to train in the fine points of herding. If a particular animal does not meet this criterion, I do not keep him for myself. This is not to say that such a puppy will not make a fine herding dog, but he will take more effort to train as he develops. I want a dog that takes the least possible effort.

Six

Learning

The late Patty Ruzzo, a leading national obedience instructor and competitor, has made a number of observations about dog training. I have not forgotten her valuable lessons.
- When teaching the dog any new exercise, the best working location is a familiar, distraction-free area, where dog attention is easy to come by.
- It is the trainer's responsibility to show the dog in a clear manner exactly what is expected of him. Commands should be given calmly and distinctly.
- A dog can only learn one thing at a time, so break down the more difficult exercises into simpler parts. Keep in mind that the dog's attitude toward a particular exercise is usually formed at its inception, so patience in this teaching phase is a must.
- It is best to work the dog in short, up beat, you-can-do-it sessions, hopefully leaving him wanting more. If the dog is confined (or otherwise away from the handler) before each lesson, he will probably be more inclined to please, and the handler's praise will be more meaningful…
- Anticipation is the first sign of learning. Do not come down too hard on the dog when he is simply trying to please.
- Train your dog when you are feeling enthusiastic, energetic, and able to give him your total concentration.
- Do each mini-exercise until the dog is performing his part happily and actively, then repeat it three (3) more times with enthusiastic praise to set it.

(Ruzzo: n.d.)

There are three phases to teaching a dog. First the dog must learn what a command means. Second he must learn to properly execute the desired behavior. (Both of these phases will be discussed in greater detail later in the chapter.) And finally, he must do so with the correct attitude. For example, you send your dog into the pen to take out the sheep. Although he performs the act (being sent into the pen) correctly, his attitude is something like this: "I'm going to bite the sheep when I get in there." In other words, your dog is too close to that fine line between a controlled and uncontrolled prey drive that I discussed in the last chapter.

In most situations dogs must be confident in performing their commands before you can get the right attitude. In the initial stages of learning a dog cannot always tell the difference between "No, don't herd," and "Yes, herd, but not with that attitude." Demanding the correct attitude too soon can sometimes make a dog refuse to herd at all, but there are exceptions.

1. You must not allow your dog to begin training with stock until his attitude is correct. For example, as you drive up the road toward the training field your dog begins barking and jumping in excitement. You are thrilled that he loves herding lessons, but his attitude is inappropriate. Here's what you do. Stop the car and insist that he change his behavior. After all, if such behavior continues it can cause an accident.

2. You arrive at the training area and your dog drags you in the direction of the sheep. Again, you are glad he wants to work, but this behavior is also inappropriate. He must learn self control before entering the training area, which in this instance means walking calmly at your side. Otherwise, he will come into the area out of control and soon bite the sheep. Consequently, you must change his inappropriate behavior right away.

One method is to shock the dog by making a quick jerk on the leash while reversing direction, thus balancing his prey and pack drive. The dog first receives a shock from the leash. His behavior is then further modified by reversing direction, in the process taking him away from the stock. In effect this is a "time out." (See page 119 for a discussion of shocks.)

Of course it's best if your dog learns correct attitude from the time you first bring him home. From the very beginning make sure your dog earns his privileges and that you carefully oversee his daily activities. He will not be resentful of this control, for it will become his way of life. For example, my dogs start by living in a crate. When they graduate from there they move into a small room. Then I allow them to stay overnight in a larger room. Finally I give them free access to three rooms—but only under supervision. They do not have access to the entire house. When I leave, the dogs are placed in kennels.

My dogs never demand anything from me. I teach them to wait for everything. They wait for their food, to go out the door, to go through gates into pastures and they wait to get into my car.

My dogs are not with us unless either my husband or I are willing to stop whatever we are doing and correct them if necessary. If we are not willing to do this, then the dogs are returned to their crates or kennels. For example, when friends visit our house we want to focus our attention on them and not our dogs. So we put them in a crate, room or kennel, which leaves us in control without the need to physically oversee them.

Many dogs that come to me for training have not had this kind of structure in their lives. They have been inadvertently taught that they are in control. Unfortunately, this can take a long time to change. It took one of my students more than two years to change the relationship with his dog. Undoubtedly, this could have been accomplished sooner had the owner been more consistent. Like this person, many students are consistent in demanding the correct attitude from their dogs during lessons but inconsistent the rest of the week. As a result, their dogs still feel they have a choice whether or not to obey commands.

In contrast, I teach my dogs (but not through force) that when it comes to obeying a command, there are no options to follow; they must obey it. Interestingly enough, this makes my dogs happy, for they know their place in the scheme of things. Consequently, they approach herding with a confident, controlled attitude; but we are getting ahead of ourselves. I will discuss what to teach your dog later in the book. You will better understand how to approach teaching, however, if you understand how animals in general, and dogs specifically, learn.

Dog training has changed dramatically in the last few years. As obedience instructors have come to better understand scientific principles of operant conditioning, they have refined old techniques and have developed new ones. Those of us involved in herding can do the same.

Stepping into the experimental laboratory may be uncomfortable at first, but the rewards are worth it. Although it is not important to memorize the terminology, it is important to understand the basic principles, for they will carry us through the entire training process.

Reward Punishment and Reinforcement

As I said earlier, our goal in training the herding dog is to teach him to work balanced on the edge between his prey and pack drives. This means we must teach our dogs to hear our voices and obey our commands, and to do this we must understand our dogs' learning process. Let's start with punishment and reinforcement. Unfortunately, lots of confusion exists regarding these concepts. One good way to think about the difference between them is that reinforcement teaches our dogs to behave in a particular way by reinforcing that behavior as, or immediately after, it occurs. In contrast, punishment teaches our dogs not to behave in a particular way because of the unpleasant punishment that results when he/she does so. For example, you want your dog to lie down so you reinforce that behavior when it occurs. Alternatively, you don't want your dog jumping on strangers so you punish that behavior when it occurs.

It seems simple, but both reinforcement and punishment come in two kinds—positive and negative—and confusion often arises when we conflate negative reinforcement with positive punishment. Let's go back to our examples. We're using reinforcement to teach our dog to lie down. There are two ways we can approach the task. First, we can start with a food treat in our hand, move it to the ground and just as our dog lies down to follow the treat, we give it to him. This uses positive reinforcement to teach our dog to lie down. What makes the reinforcement positive is that we are giving our dog something (in this case something he likes).

Yet, we can negatively reinforce this same behavior. We put our dog on a lead and walk with him in heel position. Then, we quickly step on the lead, putting unpleasant pressure on our dog's neck as our foot on the lead forces the dog to the ground. After only a few repetitions our dog will learn to avoid the unpleasant pressure on his neck by hitting the ground just as we lift our foot to step on his lead. In other words, we reinforce the behavior we want by taking something away (in this case something our dog does not like—pressure on his neck). Negative reinforcement is really avoidance training. Your dog gives you a desired behavior in order to avoid something unpleasant.

This is the same method used when training a dog with a continuous stimulation electronic collar. The dog is given a mild shock until he behaves in a desired manner. The dog soon learns that responding "correctly" to a command turns off the pain. Then he learns he can avoid the pain altogether by instantaneously offering the correct behavior to the command.

Punishment (both positive and negative) is different. Let's say you tell your 10 year old son he can't watch TV until he does his homework. Then you come home only to find him watching TV with his homework still in the book bag. What do you do? Well, in the time period in which spanking was still considered a reasonable form of punishment, you might give him a spanking and send him to bed.

This method is useful for illustrating my point. The spanking, although difficult to comprehend, would have been a positive punishment. Why? It's positive because you gave your son something he doesn't like (a spanking) for behaving in an unwanted way (watching TV before doing his homework). In contrast, let's say that instead of a spanking you send him to bed with no supper. This is negative punishment. And it's negative because you are punishing your son by taking away something he likes—in this case his supper. In herding training, the time out is a popular technique based on negative punishment. The trainer takes the sheep away from the dog as punishment for undesirable behavior.

Punishment may be necessary in some situations, but it can lead you down a very slippery slope and it often has undesired side effects—aggression being only one. It's especially problematic in dog training because punishments often occur after the undesirable behavior is done. Your dog receives an aversive stimulus (something he doesn't like) after he has behaved in a particular way (and sometimes long after).

It it's often hard for your dog to understand what he is being punished for. After all, you can't explain it to him. Imagine, for example, a dog that doesn't come when his owner calls. Instead, he returns home hours later. His owner is obviously displeased and instead of being happy the dog has returned he scolds him for not coming when called.

Unfortunately, the dog thinks he is being punished for coming home. So while the owner sees scolding as positive punishment, the dog sees it as an aversive stimulus he can avoid by not coming when called. The owner has inadvertently, but very effectively, used negative reinforcement to teach his dog not to come when called.

If you must use aversive stimuli in training, negative reinforcement techniques generally work better than punishment. In any case, timing is more critical than it is when using positive reinforcement. To be effective timing must be exact for both initiating and ending the negative reinforcement. This is because as we already saw, negative reinforcement is really avoidance training. You are teaching your dog to do something you want in order to avoid something he doesn't want.

Let's look at another example. You are teaching your dog to sit and as you give the sit command you pull up on the lead attached to his collar, thus exerting unpleasant pressure on his neck. But to be effective, the pressure must be released at the exact instant the dog sits. For you want the dog to learn he can avoid the unpleasant pressure by sitting.

The traditional approach to obedience training, what I call the "show-praise-jerk" method (show them what to do, praise them for doing it, "correct" them after the fact with a jerk on the choke chain when they do not),[26] uses positive punishment, positive reinforcement and negative reinforcement in roughly equal portions. For traditional trainers, praise is the positive reinforcement (i.e. reward) of choice.

Many of the best trainers working today in the obedience and Schutzhund fields, however, have abandoned the traditional approach, and some base their training programs solely on the kind of positive reinforcement techniques developed by such sea mammal trainers as Ted Turner and Karen Pryor. Moreover, praise is only one of the many rewards their training techniques utilize.

Although this approach may be effective in training dogs for obedience competition, herding dogs must be trained using both rewards and aversive stimuli. In the obedience ring the dog's attention must be focused on the handler and the dog must be in the pack drive at all times, even while performing the retrieve.[27] Rewards work particularly well here.

---

26 This is a variation on Fisher's "tell them, make them, praise them" (1992:7)
27 See canine theory chapter for a detailed discussion of drives

Herding is different, however. Herding dogs must balance the prey and pack drives. They must focus on the livestock and not the handler. At the same time, close teamwork between dog and handler is absolutely necessary; each must be tuned to the other. As we shall see in the pages that follow, achieving this kind of balance between drives and this close connection between dog and handler requires some use of aversive stimuli, but they must be used very carefully and then only in special situations. Why is this so?

To understand the roles of rewards and aversive stimuli in training herding dogs, let's begin by more clearly specifying what obedience is. A dog is obedient when he reliably acts in a desired manner in response to a particular cue. You give your dog a cue, the verbal command "stand," for example, and he responds with the desired behavior—he stands still.

Teaching obedience, therefore, has two parts to it: First, your dog must learn what a specific cue calls for. In our example above, he must learn that the verbal signal "stand" demands a specific response, namely, that he stands still. Unfortunately, just because your dog knows how to obey a command does not mean he will always obey it. He may know what "stand" means, but he may not stand still when you give him that command. Thus, the second part of teaching obedience involves insuring your dog will correctly respond to the commands he understands.

Research by animal behaviorists clearly show that training techniques based on positive reinforcement techniques using rewards work best in teaching the first phase of obedience—what behavior a particular cue calls for. Because of the way herding works, a limited use of aversive stimuli is necessary during the second phase—training herd dogs to reliably respond to the cues they know. Dogs in the puppy stage can only tolerate gentle aversive stimuli, however. Consequently, you should be particularly careful if you must use stimuli in your training when a dog is very young.

Rewards

To understand why some aversive stimuli are necessary we must analyze the basic elements of operant conditioning. Let's begin by looking at rewards (we will get to aversive stimuli later), for they are the foundation on which learning rests. Rewards are anything that dogs want—e.g., food, touch, praise, to lie down when tired, to chase a moving object, and for herd dogs to control stock. Because dogs want positive reinforcers, they will work to get them, and we use this in teaching specific behaviors. We give a cue (command) and when our dog responds with the behavior we want, we instantly reinforce the desired response with a reward—we give him a treat, we praise him, we allow him to herd the stock.

What are effective rewards? What motivates a dog to behave in a desired manner? No single answer to this question exists. As Ted Turner points out in his training seminars, what is rewarding in certain situations, dogs may experience as punishing in others. That's why we cannot rely solely on praise as a motivator, but must vary our rewards—sometimes using praise, sometimes using food treats, and sometimes using touch, to cite just three examples. Moreover, when using food we must vary the treat, switching between, for example, hot dogs, cheese and pasta.

What makes rewards effective takes us back to canine drives. This is overly simplistic, but in general, if a dog is in the pack drive, praise and touching are very effective; if he is in the food drive treats work well; and if he is in the prey drive, fetching a toy or ball are effective rewards. For herding dogs the very act of herding livestock is, in itself, the most powerful motivator.

This brings up an important fact about rewards. Particular behaviors, as well as, for example, food treats and fetch toys, can be powerful motivators. Moreover, when particular behaviors operate in this way, they are potentially self conditioning, i.e., the behavior itself becomes its own reward.

Unfortunately, this is a two-edged sword. We can use it in designing effective training strategies, but at the same time, we must be on our guard, for we do not want unwanted behavior to condition itself. This is why Border Collie trainers do not keep their puppies in kennels where they can see livestock for long periods of time. If these dogs are housed where they can stare at stock, this behavior can become so strongly self conditioned that training them to do anything else when they are adults is very difficult.

So then, to be effective, first, the timing of reward must be exact; we must reward a dog while he is behaving correctly. Second, initially at least, reinforcement should be continuous, i.e., given every time our dog performs the correct response. Third, the trainer must choose an effective reward. This, in turn, depends on which particular drive the dog is in.

Variable Reinforcement Schedule

Once your dog has learned a behavior, switch reinforcement to what animal trainers call a random, variable schedule. This will increase reliability. For example, first reward your dog's behavior every first, forth and tenth time, then every second, fifth and ninth time and then some other sequence of times. But never repeat the same sequence. If they cannot anticipate it, dogs will try harder to get the reward. Then reward only instances of the very best behavior.

Interestingly enough, dogs become more reliable in obeying commands if they are not rewarded every time they correctly respond to cues. Obedience trainer John Fisher explains this phenomenon in terms of gambling.

> Behavioral psychologists have likened the phenomenon seen in animals, including humans, of trying harder when an expected reward is withheld, to the gambling principle—you don't win every time, but you keep trying until you do (1992:16).

The results of this sequencing technique, though, do not last forever. As Fisher goes on to say, a dog will only try harder without reward for a limited time. If he performs in the desired way and is rarely rewarded, he will stop trying (ibid).

If variable reinforcement is inadvertent it can work against you, however. For example, if you allow your dog to occasionally creep forward when he should remain stationary, you are, in effect, variably rewarding this behavior. Then it will take considerable time to make it disappear, for your dog will continue creeping forward much longer than otherwise.

Increasing the time interval between behavior and its reward will also make your dog try harder. But if you wait too long between rewards the behaviors accomplished become mediocre. Therefore, your dog should be rewarded for performing correctly most of the time and the intervals between behavior and reward should not be too long.

I should emphasize, however, that during the initial learning stages the dog should be rewarded on what is called a continuous schedule, that is to say, 100% of the time. Moreover, the reward should come as the behavior is occurring. Then, when your dog reliably responds at least ten out of ten times move to a variable reinforcement schedule.

Shaping

At this point, we still don't know how to teach Spot to behave in a particular way in response to a given command. Obviously we cannot use words to explain how we want him to behave. One way to teach him the meaning of commands is by skillfully applying the principle of shaping; it is one of the keys to successful positive reinforcement training.

Shaping is based on the premise that:
1. Behavior is a process we can break into sequentially occurring pieces; and
2. each piece must be taught separately in its sequence.

The first step is obvious—we must identify the behavior we are trying to teach and choose a cue for it. Let's suppose we are teaching Spot the "out" command. On hearing the verbal cue "out" we want Spot to move away from the stock in a direct line. After we chose the cue and identify the desirable behavior, our next step is to break that behavior into its sequentially organized parts. "Out" has the following behavioral sequence. First, the dog looks back over his shoulder. He then pivots on his back feet 180 degrees. Finally, he walks/trots in the direction he is now facing.

Our first job is to teach Spot to look back over his shoulder when we give the verbal command "out." How do we do this? I should note that positive reinforcement can't work unless a behavior is already occurring. Then we can reinforce it regardless of how sporadic it is (Pryor 1984:23). Therefore, let's initiate a play session away from the stock. In the course of the session whenever Spot happens to turn his head over his shoulder we instantly name the behavior "out" and reward it with a treat and lots of praise. After a while Spot will turn his head upon hearing the command, "out" to get rewarded. When he responds correctly ten out of ten times (Fisher 1992:15) it is time to move to the next behavior in the sequence—pivoting on his feet 180 degrees. We no longer reward Spot for turning his head over his shoulder when we give the "out" command. Now he must pivot on his back feet. However, we might have to shape pivoting itself, so, when Spot turns his head and pivots, say only five degrees, we instantly reward him. Then, when he does this reliably (ten out of ten times) we only reward him when he pivots ten degrees, then twenty, then fifty, then a hundred. At last, we only reward Spot when he moves his head back over his shoulder and pivots on his back feet the full 180 degrees.

Finally, we are ready to teach the last phase, walking/trotting in a direction opposite from the one he originally faced. We teach this part of the sequence in the same way we taught the others, but if Spot is not getting the idea we must go back to the previous phase. It is important that our dogs are able to earn rewards or they will quit trying. Then we have to figure out how to break the new section into its subsections and shape these.

Trainers sometimes teach behavioral sequences backwards. They teach the last element in the sequence first, then the next to the last element, etc.. This is a well-known technique in language teaching. For example, in learning to pronounce a difficult three syllable word in a foreign language it is often easiest to begin with the last syllable, then the middle syllable added to the last, and finally, all three syllables together. Keep this in mind if you are having difficulty; turn the sequence around and try teaching it backwards.

Bridging

Unfortunately, there is still a question we face in all of this. Once your dog has learned all the behavioral sequences for the command where do you position yourself? Let's go back to the "out" command. You must be directly behind your dog or you cannot reward him instantly. That, however, will only teach him that "out" means come to you. But your dog must pivot 180° and go in the opposite direction no matter where you are when you are giving the cue. If you are located, for example, ten feet away on the side, how do you instantly reward him?

This is the same kind of problem sea mammal trainers face all the time. Imagine training dolphins to explode out of the water in spectacular leaps. First, you must be able to reward the animal at the exact right instant. But, how can you do this if the dolphins are swimming in a deep pool of water some distance away? As Karen Pryor explains in *Lads Before the Wind*, you must establish a signal (what trainers call a "bridge") that means the reinforcement is on its way. Then, whenever they perform in the desired manner they instantly get the signal, followed later by the reinforcement itself. Thus, such signals are often called secondary reinforcers. The food treat is the primary reinforcer in this case. Pryor used a blast on a police whistle for her secondary reinforcer because it is loud, clearly heard, and not easily confused with anything else (1975:12), but any sound (or even word) that meets these criteria will work. Today many dog trainers use a small clicker to produce such a sound.

Of course you must first teach your dog the meaning of the signal. Pryor taught her dolphins this by putting all training on hold for a few days. During this time when the animals were fed their ration of food the feeding was instantly paired with a blast of the whistle. Then when food and whistle were completely linked in the animal's mind, the next step began. Gradually, in seconds, the whistle was made to precede the food by a very brief instant, then a brief instant, and then a set of ever longer instances. Finally, the animals were conditioned to expect the reward at the sound of the whistle, even if the food came much later (1975:13).

In essence, conditioning had turned the sound of the whistle into a motivator itself. Although whistle conditioning was not an easy task (see Pryor 1975:14-15 for a discussion of the problems and her solutions to them), once accomplished, training could begin in earnest for the desired behavior could now be instantly rewarded with a blast on the whistle, regardless of the trainer's location.

So we use positive reinforcement to motivate our dogs to act in a desired manner in response to a given cue. We can use this to get rid of some simple problem behaviors. First, we identify the problem behavior and then teach behavior that is incompatible with it. Suppose, for example, your dog likes to chase livestock. First teach him the "stand" command. Then, whenever he begins to chase the stock, give him that command. Because standing is incompatible with chasing, he cannot act in the undesirable way.

Aversive Stimuli

Okay, you have used the various tools of positive reinforcement training to teach your dog the basic herding commands off-stock, and he performs them ten out of ten times. Now you think you are ready to introduce the stock. So you take your dog into a small enclosed area with sheep and say, "stand" …and your dog acts like he's gone stone cold deaf and cannot hear your voice!

What happened here? To answer this question we must go back to canine behavior. The herding instinct is, so to speak, "a wolf in sheep's clothing." As we saw earlier, the wolf's hunting behavior has been shaped and molded through thousands of years of selective breeding, so to your dog, herding is very much a special kind of hunting. Consequently, when you introduced stock he immediately went into prey drive. Thus, the kind of rewards you previously used to motivate behavior meant almost nothing to him. Getting at the stock to control them was what he cared most about. Yet, you asked him to stand, an action that contradicted his drive to get at the stock. It went against what every cell in his body screamed at him to do. It is not surprising he ignored your command.

This does not mean your off-stock work was useless. Your dog has not forgotten the meaning of the commands. You simply have to find new, more effective motivators. This is where a skillful blending of rewards and aversives is necessary. To understand why, let's go back to our example and study it further.

We understand now why the dog disobeyed, but we don't understand how he experienced the act of disobedience itself. Remember, the act of herding itself is a very powerful motivator for a herd dog in the prey drive, so when your dog disobeys "stand" (or any other command, for that matter) and continues herding, he is getting a classic positive reinforcement lesson. Unfortunately, what he is learning is not something we want to teach, for we are, in effect, teaching him to disobey. This is because his disobedience is being reinforced with perfect timing by a powerful reward—herding the stock. This is an example of accidental reinforcement (a common occurrence in animal training); you have reinforced this behavior unknowingly. If we allow such accidental reinforcements to become imprinted we have a serious problem on our hands. For this is one way our dogs learn "bad habits" and, as Karen Pryor explains, "once an animal has learned to do something, it will never unlearn it. You can overlay the learning with new information, you can extinguish the behavior almost entirely, but you can never completely erase what has been written." (1972:113). Consequently, we must intervene, and the most effective intervention is the skillful application of an aversive.

Remember, negative reinforcement works better than positive punishment and unlike punishment, always allows your dog to avoid the consequences of unpleasant stimulus by choosing to act "correctly" (as you have defined it). Take the following examples:

1. You are driving in the car with three of your children in the back seat. They are making a huge pain of themselves, as only children can do. Finally, you have had enough and say, "If this behavior doesn't stop, you're all going to get a spanking when we get home!"

    Nevertheless, the nagging, whining and picking on each other goes on. Now you say, "O.K. that's it!" Eventually quiet descends, and the rest of the trip you have three little angles on your hands. Still, you must administer the spanking or your credibility will be undermined. This is punishment. You spank your children to stop unwanted behavior from occurring in the future. Moreover, your children can do nothing to avoid the spanking. They can only grit their teeth and take it.

2. Your teenage son lives to play football. One grading period he does poorly in his academic subjects. So you tell him he can't practice with the team or play in any games unless he studies hard enough to get decent grades. The only way he can avoid this unpleasantness is to study with purpose. Here you reinforce the behavior you want (studying with purpose) by subtracting something your son finds very unpleasant (knowing he will have to watch his friends play in football games while not playing himself). Remember, though, not being able to play is a punishment for getting poor grades. The aversive stimulus is the anticipation that he will not be able to play. He can avoid this anticipation by studying with purpose. Thus the anticipation is the negative reinforcement for the behavior you want, i.e. studying with purpose.

In the wild, much learning takes place through negative reinforcement. For instance, as Crabbe notes, "In natural horse society, negative reinforcement is a part of everyday life—horses use it amongst themselves to establish what's acceptable, and what's not. When you apply it to your horse, you're speaking a language he naturally understands" (1995: 113).

If negative reinforcement is skillfully used, it can be a powerful training technique. But, as I said earlier, negative reinforcement has a down side. First, too much of it can result in timid and anxious dogs. This is because it causes animals to narrow their behavior through fear of the aversive stimulus. Moreover, even though initially it is very effective, unfortunately, it loses this effectiveness through time. In contrast, positive reinforcement is less effective in the beginning, but increases in effectiveness through time. In his training seminars, Ted Turner explains it like this: Positive reinforcement puts money in the bank while negative reinforcement takes it out.

Using negative reinforcement effectively involves three techniques: First, separate yourself from the unpleasant stimulus as much as possible, making it appear that it is not administered by you. Second, stop its application the instant the undesirable behavior ceases. Finally, the instant your dog acts in the desired way, switch to a reward, thus pairing positive and negative reinforcement to keep your bank account in the black.

Let's get back to our example. Spot has ignored the "stand" command and refused to stop herding the sheep, so we must intervene with an aversive—something he finds unpleasant. We have to make sure, however, that Spot can end the unpleasantness by acting "correctly"—i.e., in this case by standing still. The human voice is particularly effective in this situation. Some handlers find that making a sound emulating a mother dog growling at her puppy is very effective. We can use body pressure—"getting in his face," or holding him with our hands to make him stand still.

If none of these strategies work try negative punishment. Snap on Spot's lead and take him away from the sheep for a "time-out" period (two to five minutes). When you reintroduce the livestock you may find the unwanted behavior has disappeared.

More likely, you will find the unwanted behavior continues, but with less intensity. This technique can help a dog learn to behave properly, but you must have patience, for it takes time.

Some experts disagree on the effectiveness of negative punishment. Murry Sidman strongly argues against it (1989:39-40). John Fisher, however, feels quite the opposite. According to Fisher, "We need to interrupt the learned and self rewarding behavior, switch the dog's attention back to the owner and then show it an alternate behavior…" (1992:23). and, he argues, the best way to do this is to remove the reward so we can engineer new behavior (ibid). This strategy is effective because it utilizes what animal behaviorists call "extinction". Laboratory experiments have repeatedly shown that if a behavior is not reinforced (either positively or negatively) it will become extinct.

Finally, we can use various devices as aversive stimuli. Choke chains, Trainer's Edge collars, pinch collars, and electronic shock collars are a few obvious examples. But because an inexperienced trainer can easily misuse an electric shock collar I recommend the use of electronic shock collars only by experienced trainers. Consequently, I won't say anything more on that subject here.

Shocks

Here we do need to say something more about collar "corrections." Kevin Behan in *Natural Dog Training* makes the point that "correction" can be a misleading word. Our object is not always to correct the dog's behavior, but to shock it from the prey drive to the pack drive. In other words, when we give a command, we want our dog to listen and obey, rather than only focus on the stock. After all, training is ultimately aimed at bringing the two drives into balance.

Hence, the timing of shocks is different from the timing of traditional corrections. Anticipation is the key. As Behan explains, "The handler should always be thinking ahead and anticipating what the dog might do next" (1992:114). For example, shepherds in Central Europe carry staffs with them at all times. At one end is a small shovel. Shepherds use the shovel to collect dirt or sand to toss at the unsuspecting dog when he loses focus. This startles the dog, shocking him back into the pack drive.

Thus a shock is a jolt that catches your dog by surprise. It must be administered with lightening speed before Spot disobeys and as soon as we see an overly focused prey drive developing. Moreover, we should understand what the shock is for. We are not shocking Spot for being disobedient. On the contrary, we are shocking the mindlessness that is about to disrupt the balance in him between the prey and pack drive. As Behan asks:

> Is it best to react to a dog's behavior or is it better to take the initiative and ensure that the dog always performs appropriately? Why wait for negative behavior to express itself? (1992:115).

Thus, shocks are not used to punish Spot or to correct him. On the contrary, they are used to positively connect him to you. For a dog experiences a shock as vital information. "It is a pulse of dynamic energy that he will emotionally convert into a mood of attraction toward his handler." (Behan 1992:204). If done correctly, shocks attract dogs to their handlers, and help them learn to balance the pack and prey drives.

Basic Obedience

Herding enthusiasts often overlook the need for a solid foundation in basic obedience skills. Consequently, their progress is hindered by an inability to control their dog and thus their livestock. In contrast, those who develop obedience skills prior to working with stock own a considerable advantage.

A good dog on the trial field must exhibit herding instinct. Nevertheless, a dog without the willingness to instantly obey his handler's commands will not succeed on the trial field or at home on the farm regardless of instinct. A dog's willingness to follow instructions is as important as his herding ability and unless his handler can control him, a dog is of little use no matter how much natural herding talent he possesses.

The question then is, when do you begin instilling the control you need? Some argue that training herding dogs before introducing them to livestock is risky for it can turn dogs with instinct into dogs that act like remote control robots, unable to think for themselves. I have seen many of these dogs competing in trials and in predictable situations their flawless response to their handler's commands looks impressive. When the stock turn unpredictable, however, and the situation demands independent actions these dogs cannot meet the challenge.

Yet, at the same time, some pre-livestock training is necessary, for it develops a basic understanding of commands. Moreover, if the dog does not learn basic obedience the training required when livestock are introduced may shut down the dog's interest in herding. Also, pre-livestock training can at least partially compensate for minimal talent. According to Vergil Holland, just because a dog has little innate herding ability it does not mean you should not work him as long as he has the desire to work. He goes on to say:

> You should be aware that a lot of obedience will be required to compensate for the lack of natural ability. This means that the dog will need more

"push-button" work, in which he relies on your commands to help him work the stock efficiently and effectively (1994:164).

The earlier you begin obedience training the better. The handler who imprints obedient behavior in the young dog will have an easier time controlling the adult. It is not always possible to start with a puppy, however, for many enthusiasts are introduced to herding when their dog is already an adult. For those of you in this category, I suggest you start training your dog as if he were a puppy because undoubtedly he has at least some bad habits that must be corrected.

Your puppy must mature physically and mentally to out-run and control livestock before beginning on-stock training. For most dogs this happens sometime between 10-14 months of age. But while waiting for your dog to mature you can still help him develop his capabilities to obey commands. A handler who emphasizes control during this time will also establish a strong bond with his dog. One or two 10-20 minute sessions a day is sufficient.

The first herding book I read was Pope Robertson's *Anybody Can Do It*. The first few lines of the book make this point especially clear:

> The basics can be taught around stock, but in my opinion, they are easier, and more beneficial if taught away from the stock. Basic obedience is simply getting your dog to listen and respond to you instantly. Once you do take the dog to stock, it will take a little time to get him to respond as quickly as he did away from the stock, but it will come back to him very fast. Take it slow and easy. Never work a young dog more than a few minutes at a time. Quit before the dog gets bored (13).

I have trained many dogs with this in mind and all my dogs have pre-stock training. I start out in a quiet place, away from distractions. Once the dog is reliable on his basic commands, I introduce distractions—first people, then other dogs; the last distraction is the livestock itself.

General Tips

Although we are discussing foundational training here, it is also important to think ahead toward trial competition. Thus, in training any exercise, train to one level higher than you would expect of your dog in competition situations. Unfortunately, the quality of the performance of most dogs tends to decrease when you change sites. At trials, where there are lots of dogs, livestock and people present, this tendency only increases. Keep this in mind while you are working on foundational training and demand more from your dog in normal situations.

Some trainers will not take their dogs away from home until they are fully dependable. They believe it is better not to take a beginning dog to training clinics because young dogs find it difficult to work away from home. It is even more difficult to learn something new at a strange site with many other dogs and people watching. Therefore, think about attending clinics yourself; then use the information you learned in your own training sessions at home.

It is important to remember that there are many ways to achieve the same end result. If one technique does not work, try varying the exercise. Also, talk to other handlers and trainers and ask them for ideas. All dogs are unique individuals. Consequently, trainers must adjust their program to fit each dog. Moreover, all handlers are unique as well. Thus, consider all approaches and then develop a training program that works for both you and your dog.

There are certain times when you may need to take your dog back to the basics, generally when you want to increase the quality of his performance. You may also have to go back to the basics after a long lay-off period. Sometimes even adults need to return to basics. Take for instance downhill skiing. You never forget how to do it, but you need to hone your basic skills the first time you go out at the beginning of winter.

Always look ahead to the next step when training. You must be prepared to immediately advance your training program the moment your dog masters a particular step in the training process. Many times we work on an exercise for days and nothing changes. Then, suddenly the dog leaps forward in his mastery of the exercise. You must be ready to adjust your training program accordingly. This means you must plan well in advance.

Since many tending trials last at least an hour, your dog must be able to work at least that long. However, it is best to build endurance separately from learning exercises. Thus, if your dog has done an exercise correctly a few times stop training that exercise even though you may have paid for more time on the sheep. Don't forget, the last behavior accomplished is the one your dog will remember. Therefore, always try to end on a high note and quit while you're ahead.

Basic Commands

Many handlers wish to teach commands in the language of the country where the breed of their dog originated. Consequently, I have included the basic herding commands in six different languages (see next page).

English	German	French	Dutch	Italian	Polish
Heel	Fuß	Au Pie	Volg	Piedi	Noga
Come Here	Komm Hier	Viensici	Kom Hier	Vienni	Chodz Tu
Stand Stay	Bleib Steh	Bouge Pas	Blÿf Staan	Stai	Stoj
Out	Aus	Dehors	Uit	Via	Wyjdz
Jump	Hup	Sauté	Hoog	Spingeri	Skacz
Go On	Geh Weiter	Allez Va	Ga Verder	Andiamo	Dalej
Furrow	Furche	Frontière	Grens	Frontièri	Bruzda
Slow	Langsaam	Doucement	Langzaam	Lento	Powoli
Over Right	Über Recht	A Droite	Haar Rechts	A Destra	Na Prawo
Go To The Corner	Geh Ecke	Allez Acote	Ga Hoek	Angelo	Do Rogu
Switch	Rüber	Changer	Over	Cambia	Zmien
Go Out	Vor Aus	Allez passer	Ga Uit	Passerete	Naprzod
Walk On	Marsch	Marche	Loop op	Marcia	Idz

Giving Commands

The way you say commands communicates as much to your dog as the commands themselves. For example, if you say a command swiftly the dog should react quickly; if you draw the words out the dog should respond slower. Thus, "go on" means go further fast or go further slowly, depending on how you say it.,

Herding dogs have the ability to hear a command from a great distance. Although a person can hear people giving commands at roughly one hundred yards away, a dog can hear the same sound at more than 440 yards. This means you can say a command softy and your dog will still hear it. Speak to your dogs in the same helpful, patient tone you speak to a person asking directions.

How does your dog interpret what you say? According to Stanley Coren in *The Intelligence of Dogs* low pitched sounds indicate threats, anger, or aggression, high pitched sounds indicate fear or pain, and less sharp sounds mean pleasure or playfulness. This is also true of humans. When we are angry our voices drop; when we are afraid our voices become high pitched and shrill. We tend to sound sing-songy when we are happy and when we repeat sounds fast it indicates excitement and urgency. Thus, if a moving command, for example, "walk on," is said in a deep voice, the dog may not respond by moving forward because the negative tone of voice overpowered the command. The dog heard, "**DO NOT** walk on" instead of, "**YES**, walk on."

Positive and Negative Commands

Any command can be said negatively or positively. Motion commands should always be said in an uplifting-positive tone. Initially, "outs", "stop" and "slows" can be said in a quiet, firm, negative tone but at some point you must give all commands except corrections in a positive tone. For example, if a dog is always commanded to "down" with a negative tone he will assume that downing is a reprimand. But your dog needs to know that "down" is a command like any other. It is only a reprimand when he has done something incorrectly.

Puppy Training

As we saw in the last chapter, obedience is carrying out orders or instructions, and to do this the dog must be submissive. In other words he must work in pack drive. John Holmes has this to say about submissiveness in *The Farmer's Dog*:

> The submissive instinct gives us what we call a biddable dog, which can be kept under control. The stronger the herding instinct, the more does it become necessary to have an equally strong submissive instinct to balance it. (1984:27)

I base training my own dogs on this premise. In their first year I walk quietly with them through pastures filled with livestock. During this time they also learn to come when called regardless of distractions. Finally, I teach them to stop. (I'll discuss how to teach these commands in the pages below.)

If a dog does not obey these basic commands at home or in the obedience ring, he certainly will not listen in the presence of stock. *There is no stronger distraction for the herding dog than livestock.* Even when you have set the proper foundation for herding training your dog will need an adjustment period.

During this time you should proof[28] the commands that the dog already knows. For generally the lack of a timely response to a command away from the stock does not have the same consequences it has when herding. After all, your dog will eventually come when called. When your dog is chasing a lamb and does not stop on your command, however, the consequences can be serious. Veterinary fees are expensive; a dead lamb is tragic.

---

28 See page 134 for a detailed discussion of proofing

Learning to obey these simple commands has an added bonus. Herding dogs need confidence in themselves in order to control livestock. Interestingly enough, learning to be obedient dogs builds confidence, especially when working at a distance from the handler. And whether or not a particular dog turns out to be good at herding, I feel confident he can be placed as a well mannered pet.

During this same time your puppy should become accustomed to the sight and smell of livestock. So when he is seven weeks of age allow him to accompany you into livestock pens at least once a week. Your puppy will probably eat and/or roll in livestock droppings or he may notice one of the animals and walk after it.

Continue bringing him to the stock until he shows a strong interest. This will probably occur sometime between 8 and 16 weeks of age. Herding instincts surface at different times. But, you must keep the puppy away from the livestock if he shows any sign of fear, especially during his ninth week (see page 81). When your puppy has passed through the fear stage, begin taking him into livestock areas again. Once his herding instincts are imprinted, stop bringing him into the barnyard. If your puppy is having a great time chasing the stock around the entire time he is there it is time to let him grow physically and mentally before beginning serious on-stock training.

Now is the time to begin his obedience training: Your first job is to create a secondary reinforcer. As you remember from the last chapter a secondary reinforcer (or bridge) is a signal that means to your dog a primary reinforcement is coming. I begin creating a secondary reinforcer when my puppies start eating solid food (3-4 weeks of age). Each time food is placed in front of them I give the signal, "pup, pup" (although any signal will do). Each time the puppies are fed they hear, "pup, pup." In time, when the puppies hear the signal, they begin looking for food. This is the same technique used to condition the sheep to come when called (See sheep conditioning in the next chapter).When the food and the signal are linked in my puppies' minds I begin the next phase, I say, "pup, pup," let a few seconds pass and then give the puppies their food. Gradually I extend the time between signal and reinforcement until finally the signal becomes a reinforcer itself. Interestingly enough, I can use this signal as a recall command. For example, if a puppy is playing a few feet away from me I can call him to me with the words "pup, pup". This gets his instant attention and he comes eagerly for the food reinforcement he knows is soon to follow.

Free Time On a Lead

In the early stages of their training all my puppies (as well as adult dogs) drag a light cord or long line when they are out of their pens. The line insures that my dogs know they are always under my control. I want them to believe their only option is to obey my commands; they are never given a chance to experience disobedience.

At the same time, I want my dogs to run free and investigate their surroundings so they will not become timid and inhibited. Therefore, the line they drag must be light enough to be forgotten. At first the weight of the line dragging behind them may stop them dead in their tracks, especially if they are puppies. If it is light and short, however, they will soon forget the line is there. Yet a tug can bring them under control at any time.

There are times when the line will become caught on a fallen branch or other obstacle, jerking the puppy backward and administering a correction. This is nature's way of teaching the puppy to adjust his path. But remember, if you do allow this to happen you must watch your puppy carefully to insure that the line does not injure him.

Then, when his free time is over, pick up the line and walk out to the dog. This re-establishes your contact and control. Resist the temptation to establish control by picking up the lead and pulling the dog toward you. Continually pulling on the lead until the dog arrives at your feet is a form of negative reinforcement. At this stage in your dog's development, however, you should use both negative reinforcement and corrections sparingly. Instead, use rewards as much as possible. Moreover, your only objective here is to establish control. You are not teaching your dog to come when called.

Walking On a Lead

Walking on a lead is an important skill for the tending dog. He will spend many hours on a jäger (the German word for hunting lead), leading a flock of sheep with his handler. Unlike the standard variety, a jäger lead slips over your head and lays on your shoulder, leaving both your hands free. In later exercises you will need free hands to feed the sheep while you keep your dog on lead.

You should begin by teaching a puppy less than 6 months old to walk with you for very short distances off-lead. First, position yourself next to your dog. Then, show him you have food and step forward, baiting him to follow with the food. Your dog will walk next to you without realizing it. As he does so introduce the signal "heel." In time, slowly increase the distance traveled with the dog walking at your side. Remember to reinforce his behavior by giving him small tidbits of food.

Once the puppy is comfortable walking short distances, attach a light cord to his collar and attach the other end to your belt. This leaves your hands free to feed the pup as he heels. If you are working with an adult dog use a jäger lead in place of the light cord. Remember, the lead should hang loose, for it is important not to constrain your dog with a tightly held lead. This is because your dog must have freedom to move about.

When the dog becomes aware of the cord he may become distracted by it and fight with you, not wanting to be constrained by the line. This is why many trainers use food. The dog focuses on the food in the handler's hand or mouth, and not on the lead.

Most people prefer that their dogs walk on their left side but tending dogs must be able to walk comfortably on both sides. They must learn to willingly adjust to changing situations. Thus you should teach your dog to heel on both your left and right sides. Tapping your left or right leg is a good signal for this. It easily cues the dog to the side on which you want him to walk. Your dog should not be heeling and looking up into your eyes, however, for he must stay aware of his environment. After all, later in his training he must be ready to take charge of the flock at any time and he must be tuned to both his surroundings and his handler.

The previous exercises taught your dog to walk on a line by using positive reinforcement techniques. Their use helps develop a close relationship between you and your dog, but what happens when your dog gets distracted by, for example, a squirrel or a cat? Then the kind of positive reinforcers you used earlier may not be enough to keep him tuned to you. When your dog's herding instincts take over and his prey drive overcomes his pack drive, you must use a shock to restore the balance between the two drives. (Remember, however, that shocks should not be used while your dog is a puppy.)

Begin by attaching a 10-15 foot lead to your dog's flat buckle collar. If your dog is strong willed you may need a choke chain or a pinch collar. Each of these collars will produce a stronger shock than a buckle collar.

Initially, you want your dog to move about so that he can become distracted. Then wait for this to happen. When your dog is totally engrossed in the squirrel, cat or whatever, walk away in the opposite direction from whatever he is focused on. As you move away be sure the line does not become taut. A taut line will serve as a negative reinforcer, causing your dog to move with you to escape the tautness in the line. During this exercise, however, you want to shock your dog, not negatively reinforce him. Thus the lead should be loose. Then, when the lead is almost taut, give it a quick jerk followed immediately by a quick release. This will cause a shock, hopefully transporting your dog from the prey to the pack drive.

Remember, your dog *must* always be aware of you (although he must not become fixated on you). Thus he must believe you are omnipresent, even if you cannot physically

Photograph 7.1 German handler Karl Fuller a dog attached to a jäger lead.

get to him. This is why the shock is important. It makes him believe you have control over him at all times and that he must always be tuned to you, watching your body, as well as listening to your voice.

When your dog responds by quietly walking with you on a loose lead it is time to take a break. Then repeat the exercise, each time varying the distance traveled and the break time in between. As your dog begins to understand what is expected of him he will become more cooperative. He will begin paying more attention to you in spite of distractions. Some dogs enjoy this exercise and make a game of it. They enjoy trying to stay with their handler no matter how hard he/she tries to trick them.

Slow

There will be times when your dog must slow his advance to approach the stock in a calm manner. For example, during traffic exercises your dog must adjust his speed while escorting cars past the flock. Also, during the placement before the flock exercise, his approach should be steady and deliberate. This is not unlike obedience competitions when the handler is asked to change speeds while heeling.

The easiest way to teach this is by using a six foot lead. Run with your dog on-lead. Then slow your speed, and name the behavior "slow." As you slow your speed your dog will slow his as well. As the dog begins to slow his speed give him a positive reinforcer. In no time at all, your dog will learn to slow his pace on command. You can also use this method to increase your dog's speed if you have the opposite problem. Instead of slowing your speed, move faster and name the action "fast." Again, as he increases his speed, reward his behavior with a positive reinforcer.

Come Here

This exercise is easily incorporated into the free-time-on-lead exercise. Once the lead is in your hands, call your dog with a recall command. (For example, "pup pup", or "come here"). Your dog will generally respond to the cue if you taught him earlier that it signals the availability of food. Consequently, when your puppy returns be sure to reward his behavior with a food treat.

If he does not instantly respond to the command, however, give him a gentle tug on the lead followed by an instant release. This will attract his attention to you. Then say, "pup, pup" again. This time make yourself more attractive than you did before. If he still does not respond, gently pull him to you. This takes away the possibility that he will continue disobeying the command.

As the puppy moves toward you praise him. Your inviting voice praising him will attract him to you, thus encouraging him to move faster in your direction. Once he is at your feet he should receive a food treat. However, I cannot over emphasize that your puppy must *never* have the option of not coming when called. He *must know* in his heart he has no choice; he must come when you give the command.

Earlier you learned how to create a conditioned reinforcer and how to use it as a recall command. You can easily replace the conditioned reinforcer with a more normal recall command. For example, at first say, "pup, pup, come." Then in time drop the "pup, pup," leaving the cue "come". It will take only a few sessions for your dog to learn this new association. The advantage to this is that it makes obeying the "come" command a very rewarding experience for your pup. He will always associate the act of coming when called with getting a treat. Another technique you can use with equal success is to give the cue "come" whenever your puppy happens to come toward you. Then pet and praise him for coming when called.

Remember, your puppy must never feel he has the option not to come when called. Thus, if a dog is in prey drive (for example, he is chasing a cat) and refuses to come, you must administer a shock to transport him back into the pack drive. When a dog is in the pack drive his instincts tend to draw him to his handler. To increase this tendency, make yourself as inviting as possible, at the same time distancing yourself from the shock. This will make the act of coming to you, in itself a reward. As a result, coming when called will become self reinforcing.

This is how you set it up. Start with a 15 foot line, but gradually increase the line's length until you can do the exercise from 100 feet away. If your dog is young, however, keep the distance short, say a maximum of 20 feet. Attach the line to the dog's collar and then find something that distracts him (another dog, a squirrel, whatever), thus putting him in prey drive. Let him reach the end of the line and then give the command, "come here." At the same time, give a quick jerk on the line. Remember, the jerk is not a correction, but a shocking mechanism that refocuses your dog's attention on you.

Once your dog receives the shock immediately reel in the line. Don't gather the excess in your hands, but let it fall in a pile at your feet. This will make it easier to quickly reel the line in. When the dog reaches you give him lots of praise for obeying the command. Then wait until he is distracted again and repeat the exercise.

Again, it is important that your dog does not associate you with the shock. If he does, you become less attractive to him. Thus, give a tug on the lead when the dog is so distracted that he cannot see you do it and make yourself as inviting as possible by giving him a treat and lots of praise when he comes.

Dissociating yourself from the shock can be difficult when you are working alone. The dog may associate you with the shock because there is no one else around. One way to avoid this is by using a training helper. Have your helper hold the lead and administer the shock while you play the role of the good guy. As your dog receives the shock from the lead he will turn toward your helper. As he does, call to him in an inviting manner, thus attracting him to you.

Using a helper can backfire, however, if the dog comes to the person holding the lead and not the one calling him. If this happens you must make yourself more attractive to your dog than the person administering the shock. Using praise and food to override his attraction to your helper usually works.

The key to using shocks is to make the dog associate the shock with the distraction. Some dogs can become very distracted, forgetting that you even exist. Thus the magnitude of the shock should equal the size of the distraction.

"A severe shock is the only way to neutralize the emotional value of the distraction so that the dog in the future can easily put it out of his mind. Giving the dog a small shock is not doing him any favor; it will only serve to … inhibit his character by keeping him in a constant state of conflict" (Behan 1992: 297).

Because of the potential problem in using a second handler some trainers use electric shock collars. By doing so, they can shock their dogs without involving a second handler. But keep in mind electric collars should only be used by experienced handlers, for they demand perfect timing. And never use them on puppies.

Once your dog understands "come," give the command, and the instant he turns toward you, run in the opposite direction. This makes you even more attractive to him, causing him to run after you. Remember, speed is critical when working dogs on stock. The speed at which he responds to commands will determine how well your dog performs his tasks.

Walk On

Teaching your dog this command is easy. When you take him for a walk in a meadow or in the woods many sights and smells will tempt him to forge ahead of you. Any time he does so name this behavior "walk on". But remember, your dog must have a reason to walk forward. If a natural reason does not appear try throwing a fetch toy ahead of him. Investigating smells or sights and chasing the fetch toy are positive reinforcers in themselves. Thus, obeying the "walk on" command becomes self reinforcing. That's why it's so easy to teach.

If this does not work, however, you will have to use a mild negative reinforcer. First, make a driving pole. Get a 5-6 foot length of 1/2 to 3/4 inch PVC pipe, a two inch key ring and a swivel snap. At the end of the length of pipe drill a hole and attach the ring by threading it through the hole. Then clip the snap to the ring.

Begin by attaching the dog's collar to the snap on the pole. Allow the dog to drag the pole around for a short period of time (1-2 days) to get comfortable with the pole. When he is no longer afraid of it, begin the exercise. Grasp the pole at the end and use it to gently push your dog forward as you give the command "walk on." Keep the lesson time short—no more than 2-3 minutes each session.

Photograph 7.2 Pole with snap attached.

Photograph 7.3 Pole pushing dog forward.

When the dog responds to the command without a gentle push on the pole, move to a 20 ft. driving line. First, remove the snap from the end of the pole. Then attach the line to the dog's collar, running it through the key ring on the pole. This set-up will allow you to remain standing while your dog moves forward following the "walk on" command. Pull the driving line through the ring until it is tight against the pole as if the snap were still attached to it.

Begin as before with a gentle nudge of the pole. As the dog walks forward allow the line to slide through your hands, thus allowing the dog to "walk on" at a distance. It is important that the dog "walk on" without you walking with him. If you do not complete this part of the exercise you may have a dog that only walks forward when you do so as well.

Photograph 7.4 Dog walking out away from pole and handler.

Stopping Your Dog Using the Stand Command

There will be times when your dog must stop at a stand to watch his flock more effectively. Therefore, you should teach your dog to stop by standing as well as downing. It's simplest to teach a dog to stand on command when he is already standing. Just fit the command word into what he is already doing and then reward him as he does it. Remember, any behavior, no matter how sporadic, can be taught by positive reinforcement. Every time you see your dog standing still, name the behavior, "stand". Then reinforce it by immediately giving the dog a treat or by praising him.

If your dog does not stand on his own enough times for you to effectively teach the behavior try a technique developed in New Zealand. (It is used in the U.S. to teach the same behavior to pointers.) First, either build or buy a stool approximately 16 inches high, 30 inches long and 8 inches wide. This is the best size for puppies 3-6 months old. If your dog is an adult measure the width of his chest and add two inches to get the correct measurement for the width of the stool. It is important that the width of the stool is narrow enough to keep your dog from turning around and these measurements will insure that it is. The measurements for the height and length can remain the same.

You might wonder how a stool can help teach your dog to stand. Standing your dog on the stool elevates him. This removes distractions and forces him to concentrate on standing still to keep from falling off.

Now, place the stool against a wall. This will help your dog feel more secure and relaxed and will keep him from feeling too stressed. When lifting him on to the stool approach it from the rear. Once the dog is on the stool give him lots of praise. As you hold him in position (one hand under his chin and the other stroking or rubbing his stomach) say, "stand." Do not keep him on the stool for longer than two minutes. It is very important to lift him off the stool from the back. By doing so you teach him not to creep forward unless commanded to do so.

Now that he has mastered this skill it is time to proof it. (I'll discuss proofing more generally and in more detail later.) To begin, pull the stool away from the wall a distance that allows you to walk around it while your dog is holding his position. When he can stand on the stool while you walk around it increase the difficulty of the exercise by moving the stool an even greater distance from the wall. At the same time, you move away from the stool (in increments of roughly five feet) until

Photograph 7.5 Dog standing on stool with sheep.

you are circling it from a distance of 25-30 feet. If your dog moves from his position at any time go back and gently hold him in place.

When your dog reliably stands in place with you circling the stool at 30 feet away it is time to increase the difficulty again. Move the stool to a totally new location and repeat the exercise there. Finally, place the stool in a pen with sheep. This will really challenge his mastery of the command, since, as we have so often said, livestock are the strongest distraction for a herding dog.

This is the only on-livestock exercise you should do with a puppy before he matures. You may be wondering if a puppy could he be frightened by the sheep. It is possible, but standing above them should give him a sense of security.

Proofing

Proofing tests a dog's ability to perform under a variety of conditions and, at the same time, teaches him to pass the test. For example, you command your dog to stand and stay. You do not want him to move unless you give him a release command. "Psychologists call this "bringing behavior under stimulus control" (Pryor 1984:84). When a behavior is under stimulus control, the dog will not perform the act unless the signal is given, but how can you tell if your dog's behavior is under complete stimulus control? Karen Pryor defines four conditions that will confirm whether or not this is the case (1984:86). Let's use the "stand" command as an example. "Stand" is under stimulus control when:

1. The dog stands when the cue is given.
2. The dog never stands spontaneously.
3. The dog never stands when given the cue to sit.
4. No other behavior except standing still occurs when the dog is asked to stand.

The way you achieve these four conditions is called proofing.

You can test your dog on the "stand" or on any other command by:

1. Calling out any words you can think of (except the words of the release command) in an inviting voice;
2. Having someone else give a release command;
3. Introducing distractions—other dogs, people, or by even running livestock past his nose. If he becomes distracted and moves, positively reinforce him as you place him back in the exact same location. But do not repeat the command.

Repetitions of these exercises will teach your dog to listen for exact commands, to obey only your voice and to obey your commands rather than his instinct to gather stock.

But this is only the beginning. Now you have to change the conditions even further. Earlier you proofed your dog on the stand command under various conditions and in the presence of various distractions but always when he was close to you. Now you must proof the command by changing the distance.

Remember, though, when proofing any exercise it is best to break it into its components; focus on one part at a time and for the moment let the other parts slide. Then, after your dog has mastered one part, change your focus to another, demanding now a higher level of performance. Let's go back to proofing the "stand" to show what I mean.

The stand has two main components—the act of standing still, and the speed at which your dog obeys the command. When you change to a greater distance, you may have to concentrate first on getting your dog to stand on command and ignore the speed of his response. Then, when he reliably obeys the command regardless of the distance, work on increasing his speed.

But you are still not finished. Now that you have proofed by changing the distance, proof by changing your training location. At first, expect your dog to act like he is only half trained and break your exercises into their parts once again, focusing on each part separately.

It may seem unfair to set up your dog for failure by changing the situation as we discussed above. But you need to test his ability to obey commands and you must teach him to obey in a variety of difficult situations. Proofing is an important phase in his training and to be successful you must set him up for failure. Otherwise, you cannot count on your dog unerringly obeying your commands.

Imagine your dog running free in your yard. The children playing football across the street see him and want him to join their game. So they call to him and as he begins running toward the street you see a car approaching at a very high speed. But the children also see the car and call to your dog again, yelling "Come Spot! Come!" You know, however, that he will never make it across the street in time.

What do you do? If you have proofed your dog you can give him the "come" command, knowing he will instantly come to you and be safe. If you have not proofed him you can only pray, and if that fails, you can watch him die. *Proof your dog!*

It is counter productive, however, to proof your dog under more than one set of circumstances at a time. Choose a set of circumstances and proof him under these until he obeys your commands every time. Then move on to different situations until you are confident he will always obey you no matter what the circumstances. Whether your dog is a herding, hunting, obedience dog, or a beloved family pet, proofing is an essential part of his training.

It is instinctive for a dog to chase other animals but there is nothing in his genes to make him stop chasing them before he catches and kills them. Thus, it is easy to send a herding dog after a sheep or cow because his instinct pulls him toward the animal. In contrast, a dog that is called away from the stock has no instinct that motivates his behavior. Therefore, his handler must find some way to artificially motivate the dog in order to gain the desired response. Sometimes when training your dog you will want to focus on his instincts, proofing and developing them. At other times you will want him to reliably respond to commands that go against his instincts. On one occasion you may want the dog to hold the stock in a particular location; on another you may want him to release them so they can move in a desired direction.

Alternatively, your dog might continually choose to move along the same path to the stock while you may want to send him along a different route. The dog may resist your command because it goes against his natural inclination. In any case, whether you focus on honing your dog's instincts or teaching him to obey commands that are counter-instinctive you must proof your dog on each task, exercise and command. And this takes time and patience.

Many novice handlers want to move too quickly from the initial phase of training to exercises with stock. I cannot overemphasize that the basic commands developed prior to working stock *must be proofed and proofed again*. Further, they must be proofed not only where you train your dog, but in other locations as well.

As we all have heard many times, "patience is a virtue," and you must be patient to properly train your dog. Patience here means proofing your dog, then proofing him again, and in every situation you can think of. During this proofing time you may find that he may know the commands but will not perform them consistently and concisely. However, if you proof every step of the way he will gain the necessary consistency to be a truly superb working dog, one that you can rely on and be proud of.

Chapter Eight: Pre-Livestock Foundations

Eight

Pre-Livestock Foundations

Before your dog begins training on livestock you must first design and build a training area. Although a large (7-10 acre) training area is ideal you can accomplish most training in a 100 by 200 foot arena set up as diagrammed below. This will

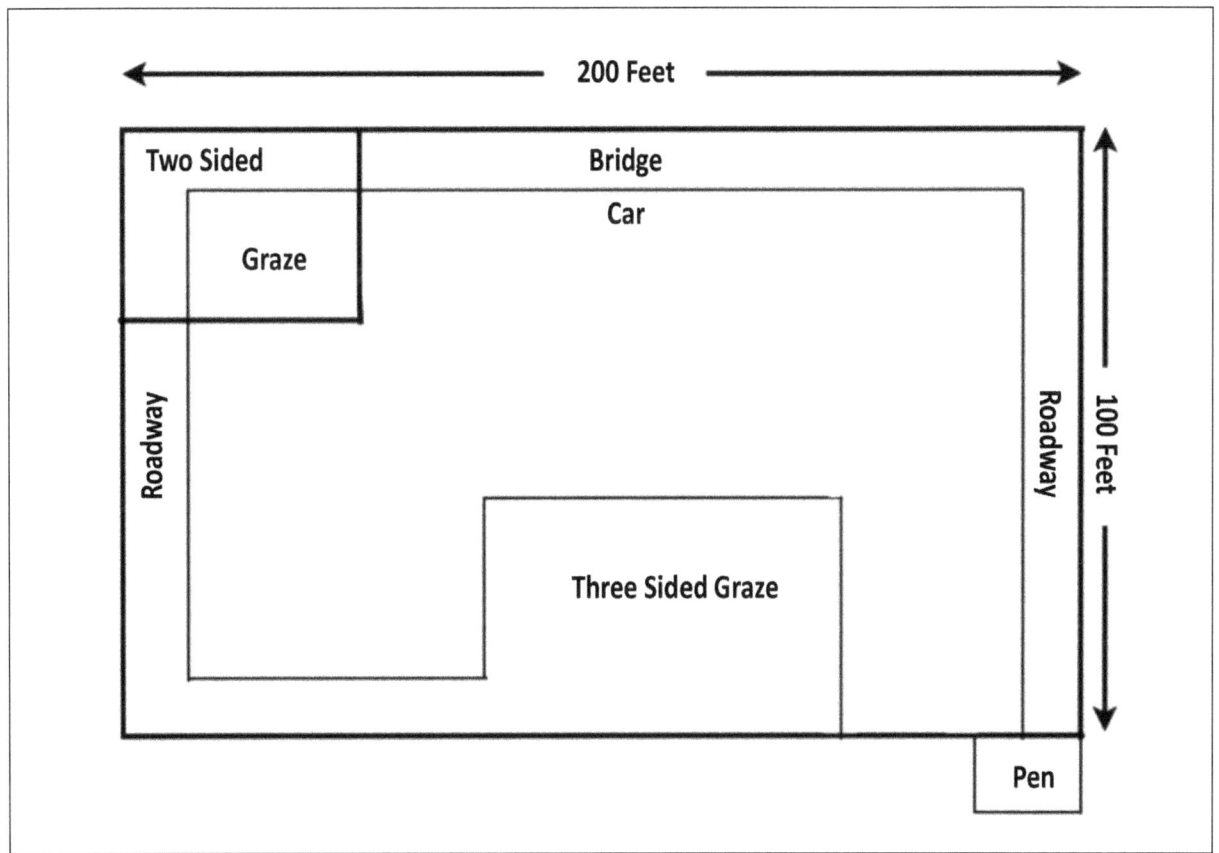

Figure 8.1

allow dogs at different levels of competency to work in the same area without the need to reorganize the training area for each dog. One roll of 100 foot plastic fencing will work well for the two sided grazing area. This can be removed quickly. The rest of the boundaries can be semi-permanent. Plow lines (furrows), ditches or strips of grass mowed short will all do equally well. The car and the bridge can be inter-changed.

Jumping

Tending dogs should learn to jump into the pen and smoothly exit the sheep. This is a difficult exercise and if you begin training without sheep you will achieve the best results. Most of the principles that underline jumping are the same for dogs and horses. I owe much of what follows to my experiences as a teenager teaching horses to jump.

When dogs run in the woods they often encounter barriers. For example, as they approach a fallen tree limb, an old stone wall or ditch, they first evaluate the situation. "How far is the limb from the ground? How deep is the ditch? What is the necessary speed of approach to clear the wall? What type of footing is necessary to clear the obstacle? Do I have the physical ability for the maneuver?" Obviously your dog does not consciously ask himself these questions. Nevertheless, his brain computes this information and his gait changes to match the stride he needs to pass over the obstacle.

It is important to remember that your dog should not jump obstacles until he is roughly a year old. And then he should only jump obstacles as high as his elbow. Nevertheless, you can teach your young dog to stride over ground poles while he is maturing.

Different breeds of dogs mature at different ages and individuals within breeds develop at different rates. Therefore, consult your veterinarian about when your dog should begin his jump training. Because some breeds of dogs are more prone to hip dysplasia than others, preliminary x-rays are also advisable.

The first rule when teaching a dog (or a horse, for that matter) to jump, is to start low and gradually increase the height until you reach your goal. Horse trainers spend many hours walking and trotting their horses over poles before asking them to attempt a jump.

There are two aspects to jumping, the stride[29] and the take-off point. First, you must teach your dog to stride properly over a pole. Suzanne Clothier in her workbook, The Clothier Natural Jumping Method, has mathematically worked out the correct

---

[29] A stride is a measurement of distance. It is the distance between the points at which the same foot hits the ground while your dog is walking. For example your dog steps out using his left foot (point A), then places his right foot ahead of his left. Next he picks up his left foot and places it in front of his right foot (point B). The distance between point A and B is his stride.

stride length using the dog's height, length, and elbow height. She classifies dogs into four body types:
1. Rectangular bodied with height at the elbow 50% of total height at the withers.
2. Square bodied with height at the elbow 50% of total height.
3. Square bodied with height at elbow over 50% of total height.
4. Rectangular bodied with elbow height less than 50% of total height at the withers.

The stride of types (1) and (2) dogs is three times their body length; the stride of type (3) dogs is body length plus 1/3 body length times three, and the stride of type (4) dogs is body length minus 1/3 body length times three (1990:30-33).

To jump properly a dog must roughly calculate the above stride measurement to insure he clears the obstacle cleanly. Many dogs do not calculate the distances correctly, causing them to jump improperly. The above calculations can guide you in helping your dog jump with correct form. You must be aware, however, that some dogs jump just high enough to get the job done but without correct form. Repeated jumping using incorrect form can cause serious injury. The type of jumping tending dogs are required to do is very challenging. Therefore, the dog must learn to jump correctly.

Begin by walking over a pole yourself. You will notice that your stride changes as you approach it. These changes are similar to the adjustments your dog makes. A dog must adjust so that he reaches full stride in order to properly clear an obstacle.

Next, put your dog on a six foot lead (the lead should be loose at all times) and be prepared to walk with him over a pole lying on the ground. Start your approach two strides from the pole using the above measurements. These measurements will place your dog at the correct distance to reach his full stride when he is directly over the pole; thus he will effortlessly clear it. If your dog does not reach his full stride when he starts jumping obstacles he risks: 1) tripping over the obstacle, hitting it with either his front or rear feet, 2) impaling himself, 3) stopping dead just in front of the obstacle and refusing to jump.

It is important for your dog to keep his focus on the pole as he strides over it. He should not be watching you, for he will trip over the pole if he does. If he begins to focus on you, you must adjust your position. Instead of running along side of your dog, construct a jumping corridor or chute and stand at the end of it.

Some dogs may refuse to jump when the handler is not moving along side of them. They need to work in a more enclosed corridor or jumping chute. You can build a jumping chute by placing the jumps between two fences, thus forming a corridor. The chute should be wide enough to allow the ends of the jumps to touch the sides of the fencing. The jumps themselves should be 4-5 feet in width. Position the dog at one end of the chute while you stand at the other. There are three advantages to this. First, it allows the dog to jump freely. Second, it allows the dog to choose his own speed. Finally, it keeps the dog from running around the jump.

In order for this to work you must motivate your dog to travel through the chute and over the obstacles to reach you on his own. Therefore, you must become more attractive to him than not navigating the chute and jumping the obstacle(s). He will willingly do this if he wants to get to you. You must coax your dog through the chute by becoming so attractive (using praise and/or food) that he wants to get to you regardless of the obstacle(s).

At the beginning of training, the walls of the chute must be parallel to one another and in a straight line. They also must be flexible, for later you will have to reconfigure the chute, first to form a single curve and then an "S" curve. Consequently, it is best to make the walls of the chute from electrified netting. This netting is available in several heights including 33 and 42 inches, and two lengths, 75 and 150 feet. It is very light weight even though posts are included with the netting. Thus, a 150 feet section can be set up in ten minutes and removed in half that time.

There are some advantages, however, in teaching your dog to jump on a lead rather than in a chute. Working on a lead does not give your dog the option of refusing to jump. On the other hand, taking him off lead and allowing him to jump more naturally also has advantages. An animal that is allowed to jump freely can adjust his stride and balance more effectively than one constrained by a line.

Be aware, however, that human error can effect how your dog jumps. For example, suppose you unintentionally jerk the lead as your dog reaches the correct take-off point. The jerk will correct the dog jumping from at the correct point, and thus make your dog unsure of where the correct take-off point is.

When introducing a jumping chute start the dog walking through the chute without poles. Once he is accustomed to the chute, gradually add the poles. After your dog has walked over one pole a few times add a second, then a third, and finally, a fourth pole. Set all the poles at ground level and far enough apart (2 strides length) so the dog can stride over them with a smooth natural gait. In time your dog will learn to automatically pace his gait over the poles without having to make adjustments.

Once your dog has mastered striding over a pole, it is time to add height. First, however, you must develop his take-off. Horse trainers do this by using an obstacle called a cavalletti. Cavallettis have three built-in jump heights, 5, 10 and 16 inches. The trainer changes the height by turning the jump over. If more height is needed you can stack cavallettis on top of one another.

Now replace the last pole with the cavalletti and set it at its lowest height. In time you will roll the obstacle over to increase its height.[30] As the height increases the dog's

---

30 Some dogs may need their jump height increased slowly in increments of 2 inches at a time. If this is so, a different type of jump must be used in place of the cavalletti. Generally the jump does not need to be changed until the dog is jumping at much higher heights.

speed must increase correspondingly. This means the last stride before jumping must also change.

When learning to jump over a cavalletti a dog (or horse) must learn to collect himself both prior to take off and during the jump. When he collects himself he brings his hind end underneath his body, shifting his weight to the rear. As he jumps his head is stretched forward, his withers are at the peak of the arc of his body's curve, and his legs are tucked tightly against his body.

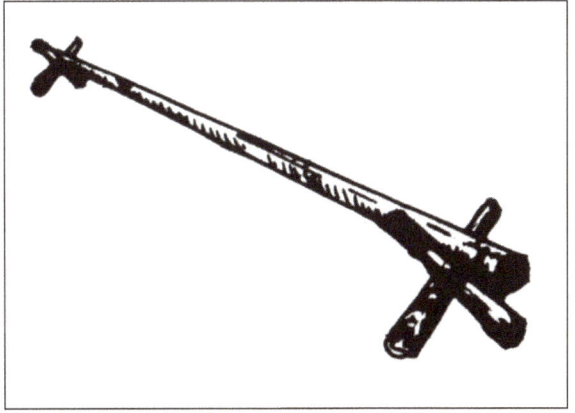

Figure 8.2 **Cavalletti**

The formation of the arc is the most important factor in jumping. If your dog jumps too soon his arc will peak before he reaches the obstacle and he will crash into the back of it. But if he jumps too late he will crash into the front of the obstacle before he is high enough to clear it. Thus it is important that your dog take off at the correct distance.

You can teach your dog to form the correct arc by using a cue for the correct take off point. This point is located at a distance from the obstacle that is equal to its height. For example, if the jump is 30 inches high the take-off point is 30 inches from the

Photograph 8.1 **Horse Jumping showing correct form (Photo courtesy of "Der Hannoveraner").**

Figure 8.3 Jumping curvatures

jump. Placing a pole on the ground at this point will cue your dog where to take off. Horse trainers call these cues ground lines. "Ground lines are essential to prevent a green horse[31] from getting too close to a fence; they also ... (encourage) him to lower his head and neck and to round his back into a "bascule" or arc" (D'Ambrosio and Price 1978: 51-52).

You will often see dogs at trials jumping incorrectly, a fault noted by judges. Because they jump too soon they must use their legs to push off the top of the fence, but they jumped too soon because their take-off point was too far from the jump. Dogs taught to jump properly will not develop this fault. It is very important that your dog learn to jump properly before beginning training on stock. He needs a strong foundation before jumping into a pen with sheep as a distraction or he will develop bad habits.

Although your dog will only be jumping over one fence as he enters the sheep pen, it is still important to teach him to jump over a series of obstacles to help him develop balance and suppleness. In observing Olympic gymnasts as they perform tumbling runs I am always struck by their grace, balance and flexibility. The tending dog must have these same skills to turn and jump into a pen properly. Without these skills your dog will run the risk of hurting himself as he does so. Remember, this is a very difficult maneuver. Your dog must run, turn, jump and land perfectly in the pen without hitting any sheep. Therefore, he needs the balance and suppleness of an Olympic gymnast.

Start with four poles spread two strides apart. As a warm-up exercise, work your dog over the poles first. Then increase the height of the fourth pole by replacing it with a cavalletti. Then increase the height of the third obstacle, then the second and finally the first.

Once your dog is jumping a series of jumps in a straight line it is time to change the configuration of the jumping chute, for now you must begin to teach your dog to bend his body before, during and after his jump. Dogs that perform obedience jumping do not need this extra flexibility because they always jump in straight lines, but because your dog must make one or two left or right hand turns before jumping into the sheep pen, he must learn to bend his body.

---

31   An inexperienced horse

The first variation will be to curve the chute to form a circle. Place the jumps two strides apart within the circular chute. This configuration will cause your dog to bend his body as he approaches each cavalletti. Send your dog in both a clockwise and counter clockwise direction to develop flexibility on each side.

Next change the chute's configuration to form an "S," again placing the cavallettis two strides apart. This will require your dog to bend more before, during and after the cavallettis, thus helping him develop a smooth circling approach to the jumps. Again, work your dog in both directions. Because the distance is greater than it is when working in a half circle, you will need more jumps.

The final configuration of the chute will require your dog to jump after making a 90 degree turn to the left or right. This configuration (a dog's leg) will help your dog develop a smooth circling approach to the fence, for if he does not approach the fence smoothly the jump will be awkward.

Because your dog will have the added difficulty of having to turn before jumping, you must lower the height of the jumps. Set each jump two strides from the corner in each direction.[32]

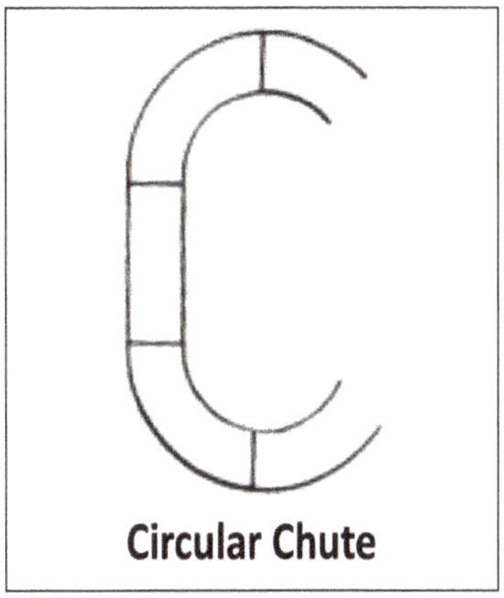

Figure 8.4 "C" Chute.

Figure 8.5 "S" Chute.

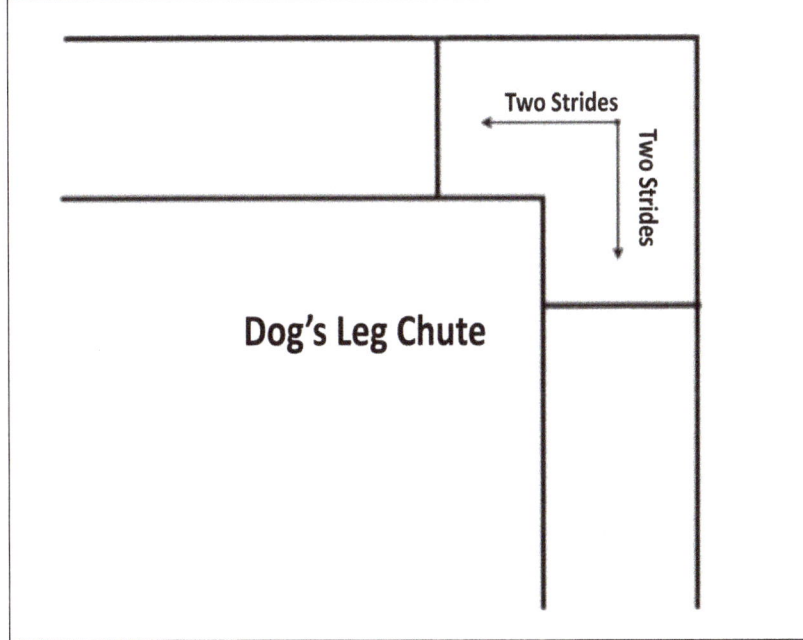

Figure 8.6 Dog's leg chute.

---

32 The second obstacle is not required, but it is a time saving device. With two obstacles, you can quickly change directions because you will not have to move any cavallettis.

143

After clearing the first pole your dog will have two strides before he reaches the corner. Once he makes the turn he will have two more strides before reaching the next obstacle. The deeper into the corner he goes the wider his circle must be to approach the fence correctly. When your dog is comfortable using the new configuration, increase the height of the obstacles.

The act of jumping is only a small part of taking the sheep out of the pen. To accomplish a good take from the pen a dog must:

1. Stand at the opening of the pen;
2. Run along the outside of the pen;
3. Jump over the fence;
4. Stand;
5. Run to the corner of the opening;
6. Stand at the opening of the gate (see pen exercise page 193).

At a trial this chain forms one fluid motion. When teaching it, however, we must break it into its elements and teach each part separately. But teach them backwards. Work on the stand at the corner first. Then add the stand after jumping into the pen. Next add the jump, etc. By doing the series backwards the dog is reinforced by the part of the exercise he already knows. This will shorten the learning time. Then, when your dog has learned all the parts put them together.

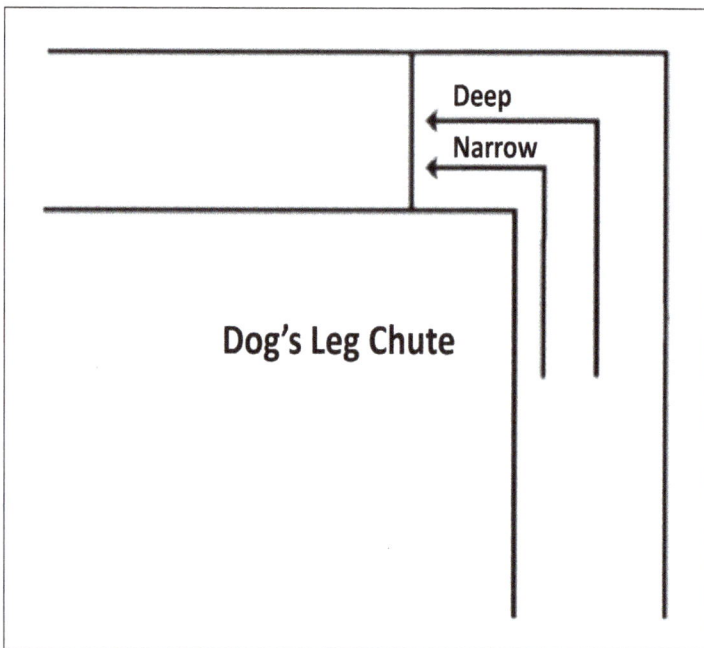

Figure 8.7 **Deep and narrow approach.**

Teaching the Out with Body Pressure

I described teaching the "out" using positive reinforcement in the previous chapter (see shaping section page 119). Positive reinforcement often encourages dogs to move toward the reinforcer. It's easy teaching a dog to go toward something he wants, but difficult teaching him to move away from it. Remember, the stock is the most powerful attraction to a herd dog. Thus when teaching the "out" in the presence of stock, you will probably need to use negative reinforcement.

Unfortunately, using the negative reinforcement technique I discuss below may cause initial problems. Until now you have been careful to make yourself attractive to your dog. In this exercise, however, you will do the opposite by using your body to exert pressure. This change by itself will require your dog to make a difficult adjustment. Moreover, the act of moving away from the stock goes against every fiber in his being. That's why introducing this particular exercise in the presence of livestock may be hard for your dog unless you have already taught the "out" using positive reinforcement. If you don't omit this step, however, and teach this command on stock using the pressure of your body, the dog may interpret "out" to mean "don't herd the livestock." Thus, introduce your dog to the command away from stock using positive reinforcement. Then when you introduce stock re-teach the command using the technique I describe below.

Some dogs, however, must be taught to deal with pressure placed on them before they are introduced to livestock. These dogs have a very strong pack drive as well as a strong herding drive. Consequently, they are very sensitive to their handler. It does not take much body pressure for them to move away from you. Unfortunately, they often interpret this kind of pressure to mean they shouldn't herd livestock. Even though you teach these dogs the "out" with positive reinforcement away from the stock, they must be taught the command using negative reinforcement away from the stock as well. Here, however, your main goal is to teach your dog to handle body pressure from you. Once he can handle the pressure away from stock introduce the technique with livestock. This will allow him time to adjust without dampening his herding instinct. Remember, though, positive reinforcement works best when teaching dogs what commands mean, so don't omit the first step in training this kind of sensitive dog.

Start with the dog on a six-foot lead. Stand directly in front of your dog and walk toward him, giving him the command "out." This will cause the dog to become uneasy. As a consequence, he will want to move away from you. His first response will be to move either to his left or right, but you want him to turn his head back over his shoulder and retreat. So, if he moves to the left or to the right, stop him with a "stand" command. This will tell him that moving to his left or right was incorrect.

When he is once again standing, move toward him and give him the "out" command again. If he moves in the desired direction (even if he only turns his head) praise him. Soon he will turn his head over his shoulder and move backward on command. At first you can be satisfied if he moves backward just a few feet. Then, slowly extend the distance until he will move backward 100-250 feet. To increase the distance, move with your dog while he is retreating, keeping gentle continuous pressure on him until he covers the desired distance. Once he reaches that distance release the pressure completely.

Now, increase the difficulty by not moving with him as he retreats. Stay where you started the exercise. If necessary, take a few steps (or lean) toward your dog, exerting pressure on him. Finally, ask your dog to "out" from many different positions.

As I said earlier, the speed at which a dog responds is important. Now that your do has mastered the act of turning and moving away he must learn to increase his speed. This cannot be taught at the same time he is learning to turn away. Remember, it is difficult, if not impossible, to teach an animal to do two things at once without confusing him.

Furrow Walking

Much of a tending dog's work is forming a living fence to keep his charges away from cultivated fields or out of harm's way. Such work is based on furrow walking and like jumping, the foundations for it are best laid away from stock.

Choose a boundary line in your training area which can either be a plow line, a ditch, a strip of grass mowed short, or a roll of plastic fencing rolled along the ground. The boundary should be at least one hundred feet in length and discernible by sight and touch. The dog should be able both to see the boundary and feel it with his legs and/or paws. In time he will be concentrating on the sheep and thus, must be able to feel the boundary line.

Picture in your mind, the sheep on one side of the boundary and your dog patrolling the other side. Start by walking with your dog on a 25 foot lead along side the boundary. If you use a furrow or mowed strip of grass, the dog should work in the furrow. At the beginning of each training session walk the dog along the boundary line to familiarize him with it. In time you will not have to walk the entire length with your dog. As you and your dog walk the boundary line, give him the command "furrow" to indicate the boundary.

As the dog bumps up against your leg, or moves his feet out of the furrow, give the dog a correction on the lead and the command "out." The dog should retreat off the boundary at the same location he came over the boundary line. Praise the dog and continue walking along the boundary line.

When using an artificial boundary like plastic fencing, the dog may not want to walk on the fencing. Allow him to walk alongside it while you walk on the plastic. In time, move across the boundary to the sheep's side of the line. Later, after you begin working sheep, the boundary will divide you and the stock from the dog.

As you reach the end of the boundary line, reverse your direction by turning the dog in toward the imaginary flock. It is important to encourage the dog to turn his body toward where the sheep will be. If you allow the dog to turn the other way, later when you introduce stock, he can lose sight and control of his charges. When using a furrow as a boundary, the dog must keep his back legs in the furrow as he turns

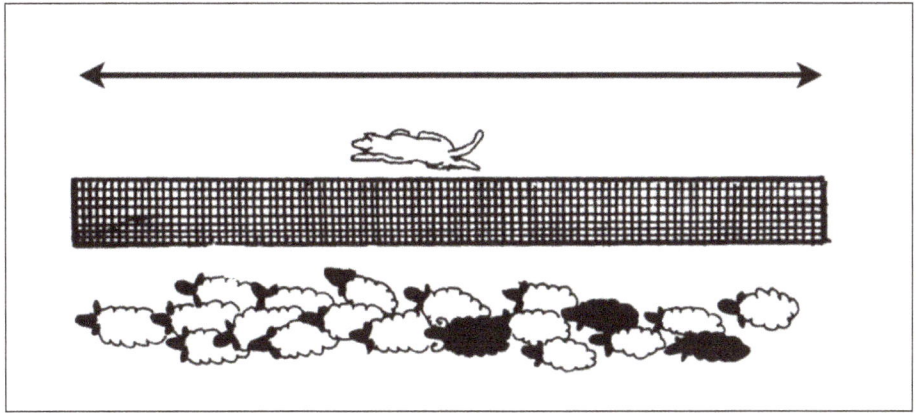

Figure 8.8

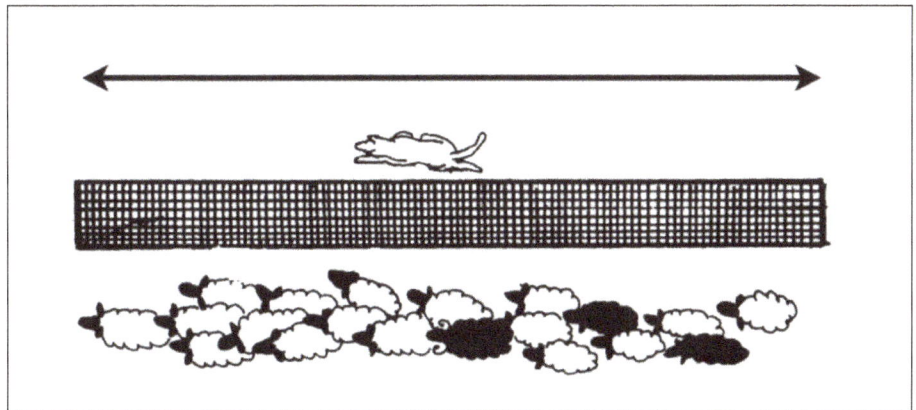

Figure 8.9

within it. This will cause the dog to roll back over his hind end, causing him to pivot while changing direction.

As the dog progresses, release more lead to allow him to move more fluidly and to move further away from you. This is an excellent time to integrate the "come here" and "go on" commands. You can use the "go on" command when the dog is traveling in any direction other than toward you (see Figure 8.11).

If you have difficulty walking the furrow while holding the lead and keeping the dog from entangling you, I have some hints that have been helpful to me and others I have worked with.

Photograph 8.2 **Malinois working in a furrow.**

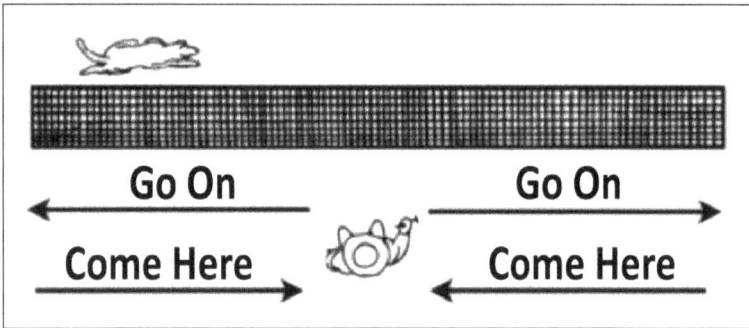

Figure 8.10

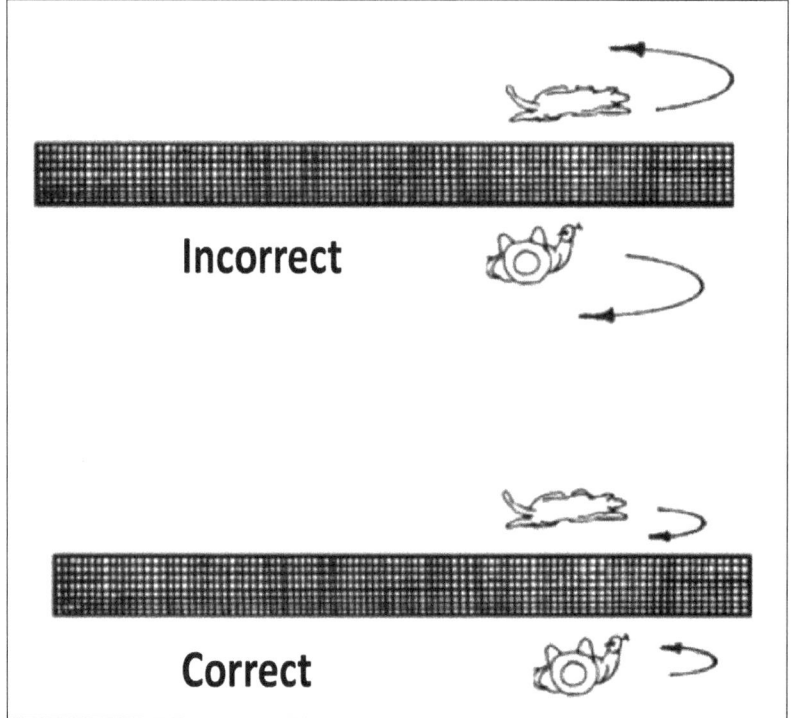

Figure 8.11

Allow the lead to drag behind you. You don't have to carry the entire lead in your hands.

When walking with the dog on your left side keep the lead in your left hand. Similarly, when the dog is on the right, the lead should be in your right hand. Just before reaching the end of the boundary line, change the lead from one hand to the other to prepare for the about turn.

Turn out toward the dog while the dog turns in toward the stock and/or you. In doing so, the turn becomes smooth and flowing.

When a natural boundary (a plow line) is not available, I suggest using plastic fencing for the following reasons.

- The fencing is very light-weight and portable. You can adjust it during training and can set it up just about anywhere.
- The fencing is easily noticeable by the dog and handler.
- Because the fencing is approximately 100 feet long it gives enough distance to practice before you must turn around.
- The plastic fencing has a width of 40 inches, which is ample for the handler to walk on. The width also helps separate you and the dog when you wish to lead the flock and control the dog at the same time.
- The texture of the fencing itself reminds the dog not to cross it.
- Even though the cost of the fencing is high, you can use it for penning the stock once the dog has started to work natural boundaries.

Photograph 8.3 Walking along plastic fencing, Photo courtesy of Kevin Udahl

Conditioning Your Flock

Now you have laid the training foundation for your dog to begin working stock, but you must still condition the stock before introducing your dog into the equation, for sheep must trust and follow their shepherd in the tending style of herding. Without this trust, training a tending dog to effectively herd sheep will be almost impossible. The sheep must follow the shepherd because they know his voice and this requires that shepherds condition their sheep.

In each flock there is a lead or dominant sheep. Give this sheep a treat each time you tend to the flock. She will then become almost a pet sheep and when she hears your voice will come for the treat (which can be corn, bread, or even a slice of apple). Conditioning a sheep in this manner gives you a tool to use in moving your entire flock. As we saw, sheep generally follow the one in front of them, not wanting to be left behind. Thus, if one moves forward the rest should follow.

It is important that you lock up your flock when they are not grazing. If you do so they will be more willing to be led to pasture, for their stomachs will be empty. Also, the flock will begin to know that when you arrive they eat.

Figure 8.12

Walk in among the flock. Call to the lead ewe and feed her. You will find that other members of the flock will begin to move toward you when you call the lead sheep. Use a one syllable word when calling the flock to you. In Germany many shepherds use, "woop, woop" (for more information see page 38).

When breaking in a new flock I use a small bucket to feed the sheep. The noise of grain in a bucket usually gets their attention. Once the flock is conditioned to my arrival I replace the bucket for a carpenter's apron. This allows me to move more freely around the flock without the clumsy bucket. It also allows my hands to be free to work with my dog. In time I replace the apron with a small amount of food in my pocket. Also, if I wear the same coat each time I visit the flock, it responds quickly to anyone wearing that same coat.

As you exit the pen the lead ewe will begin to follow you. Then, the rest of the flock will follow the lead sheep, which is right at your heels. At first the lead sheep will not be sure she trusts you enough to walk past the dog you have stationed near the opening of the gate. Thus, it is better to start without the dog present.

At first you may have the confidence of the flock for only a short period of time. Be patient. In time the flock will follow you anywhere you lead them. Once the flock will follow you for about a quarter of a mile, they are properly conditioned. You can now begin on stock training.

Nine

Basic On-Stock Training

On Stock training should begin by teaching your dog to escort the flock from the pen to the training area. Start by placing the dog far enough away from the gate to give the lead sheep the feeling of safe passage. In time, as the flock begins to relax, you can move the dog closer to the opening of the pen. A good place to start is about 20 feet from the edge of that opening. When the flock becomes comfortable with the dog at this distance, move him five feet closer. Keep inching your way closer until the dog is even with the opening of the gate.

As you walk to the training area you may experience the flock's pushing past you to find forage, leaving you in their dust. An easy solution to this problem is to walk with the dog on lead in a loose heel position. This will help control the flock's forward motion and direction.

At times you will wish you had a second pair of hands, one to hold on to the dog and the other to feed the flock. You can get the same result by using a hunting lead. This type of lead loops over your shoulders and around your body, thus giving you the freedom to work with your hands and still have control of the dog (see Photograph 7.1).

But now another problem will arise. The flock will be unsure of you because of the dog's presence in front of them. You and your dog work as a team. Both are equal partners in tending the sheep. Therefore, the flock must have confidence that no harm will come to them, that both their overseers have their best interests at heart. This means the dog must remain calm while walking with the handler. Otherwise the flock will not follow. If the dog stares at the sheep while walking in front of them, however, they will also refuse to follow.

In tending, a "loose eyed" dog works better than a "strong eyed" dog. The staring and predator-like manner of strong-eyed dogs prevents the flock from trusting them. Generally, strong-eyed dogs are used to prevent the flock from moving toward the dog.

The dog communicates that the sheep should fear him, and, as a consequence, move away from his presence. In contrast, the tending dog should communicate trust. His working relationship is built upon the confidence of the flock, similar to that which the flock has with a livestock guarding dog. Indeed, some dogs are asked to perform both tasks, to herd the flock by day and protect them by night. This is what the protection exercise in the herding competitions in Germany is designed to test.

While leading your flock the sheep may stop to graze on their way to your destination, but be patient the first few times out. Eventually they will move at a quicker pace. Once they know you are bringing them to a new place to graze, they will move more quickly, especially if you repeat your daily routine.

In the past, flocks in Europe were sometimes led by shepherd boys, donkeys, goats, rams, and/or dogs. If the flock is led by a shepherd, however, it will move at a quicker pace and the column will become narrower or more elongated. When the flock is led by an animal it chooses its own width for the column and pace and may spread out across an entire roadway.

Once the flock is conditioned properly place your dog on a "stand stay" right at the pen's entrance. The dog should stand quietly. He should not stare at the flock as they follow the handler past him or exhibit any predator mannerisms. This would keep the flock from trusting both the handler and the dog.

Watch your flock at all times to judge its reactions to the position of the dog. If the flock does not trust the dog, move him back far enough so that they will pass by without fear. The relationship built at this time will drastically affect the amount of time needed to gain the confidence of the flock. You will find that the flock has less confidence in you if your dog breaks their trust. The flock associates the dog's behavior with yours. Therefore, take the time to carefully condition your flock to the dog's presence and make sure your sheep are willing to listen to both you and your dog's voice. Now you are ready to move to the next exercise, training your dog to work boundaries.

Photograph 9.1 Flock in Germany lead by a donkey who in turn is lead by a shepherd; his dog is a Briard.

Figure 9.1

Photograph 9.2 Two German shepherds working to protect sugar beets in Germany.

One Sided Boundary With Sheep

Here he must first learn to walk in the "Furrow," that is to say, run up and down in the last furrow in front of the crop he has to protect. He is not allowed to do damage himself by stepping on the crop, and he must get to know that particular furrow is the utmost limit beyond which he must not allow the sheep to come (Von Stephanitz 1925:296).

Before now the dog's work with sheep was very controlled. He learned to relax while walking at the front of the flock with you and to stand quietly while you led the flock past him. In this exercise, however, the dog will have more freedom to move about the stock. You will also begin to proof the pre-stock work for the one sided boundary and will no longer have to imagine the sheep on the boundary's other side.

All exercises with stock begin by standing the dog at the pen and leading the sheep to the training area. Once you arrive, begin work by naming the boundary line with the command "furrow." Don't be surprised if the dog is more aware of the sheep than he is of the boundary line. If he is overly interested in the sheep, stop and stand quietly with the dog for a while. Once the dog is settled and quiet, begin walking along the boundary with the dog on a loosely held, long line. Although the long line may be 30-50 feet in length start with a much shorter distance between you and your dog (approximately ten feet). You should walk between your dog and the sheep along the boundary line. Your position will separate the dog from the flock, giving you an advantage in anticipating the movements of the dog. At the beginning of each training session you should alert the dog to the presence of a boundary.

By this time the dog should understand the command "furrow". You are now proofing him on this command in the presence of sheep. Take a position between the dog and the edge of the boundary as you did in the pre-stock section. Remember, the command "furrow" names the boundary, not the motion of the dog patrolling it (see Figure 9.2).

Figure 9.2

Once the sheep are on the other side of the boundary line the dog may focus so intensely on his charges that he will not respond to the "out" command as well as he did without the stock. This will take some time to correct. The livestock are a very strong distraction. You want the dog to concentrate on the stock but, at the same time, to respond to your commands. The dog should retreat from the boundary when given the command to do so. Because you have placed yourself between your dog and the boundary you are in a good position to use body pressure to make him move back. Sometimes, however, body pressure alone does not work. In instances like this, some of my students have found that bouncing a basketball or soccer ball directly in front of their dog has the desired result. Remember, though, your object is not to hit the dog with the ball. That would be punishing him for moving toward the boundary. Instead, the object is to bounce the ball at the dog in such a way that he can avoid it hitting him by moving back away from the boundary. By bouncing the ball in front of your dog, you are reinforcing (in this case negatively) behavior you want to occur, rather than punishing behavior that you do not want to occur. The difference may seem small, but negative reinforcement is generally more effective than positive punishment.

When responding to the "out" command the dog should move back outside the boundary along the same route he used to cross over the line. Praise the dog and continue to allow him to transverse the length of boundary on his own (see Figure 9.3). As the dog works the furrow you should see him instinctively turn in toward his sheep. This is because concentrating on the stock causes him to turn toward them. Continue to walk along the boundary line with the dog as you did before. The dog should keep its attention on the sheep, turning towards them. If he does not, encourage the dog to do so. Be sure that the dog's hind legs don't leave the furrow. Remember, if the dog is allowed to turn away from the stock he will lose control of them. As the dog progresses in his training you can release more lead to allow for more fluid motion. At this time, work on the "come here" and "go on" commands.

As you patrol with your dog, you will notice him beginning to flow back and forth. This graceful trot is the hallmark of European tending dogs. Once the dog moves consistently in this controlled, flowing manner, and once he reliably responds to the commands, "come here" and "go on," you can release him off lead. Eventually he will know to turn at the tail end of the flock and return to its head.

When each training session is completed, lead the flock back to their holding pen. Place the dog on a "stand-stay" at the opening of the gate and lead the flock past the dog into the pen.

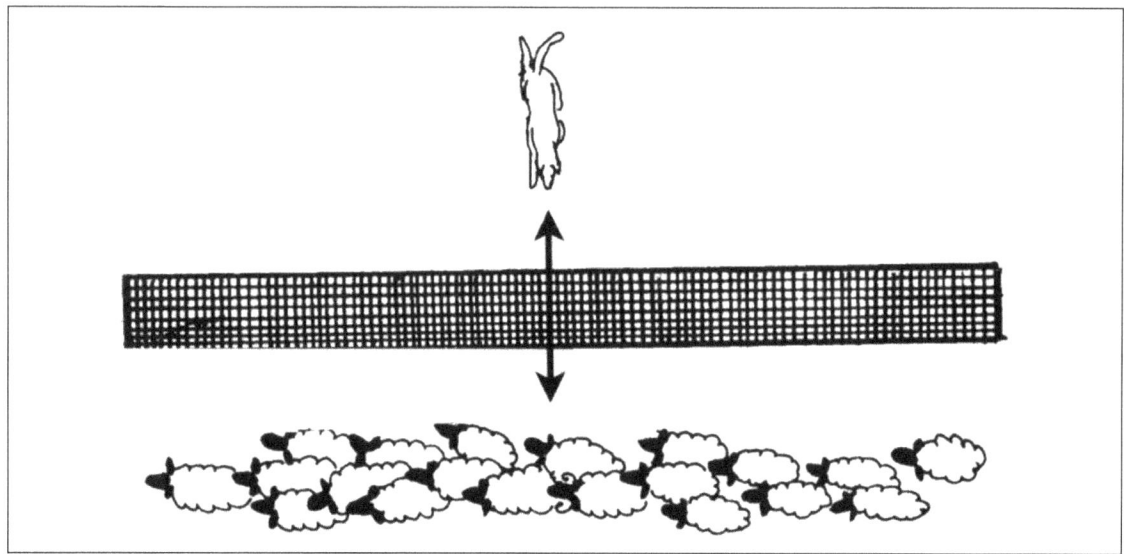

Figure 9.3

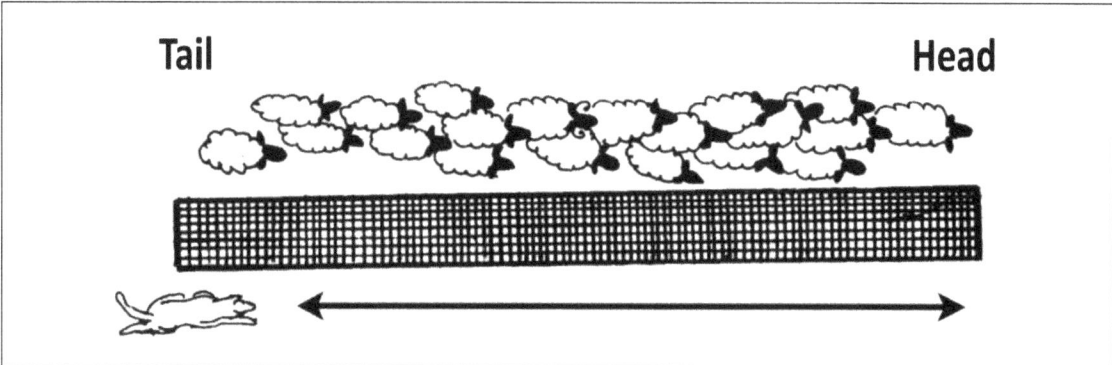

Figure 9.4

Movement of the Flock Within the Boundary

When the dog settles into a pace he can sustain for an extended period of time, i.e., when he trots back and forth under control, start this next phase of his training. Previously you conditioned the flock to move past the dog while he was stationary. Now you will condition the sheep to the movement of the dog.

Position yourself in front of the flock. This will allow you to lead the sheep forward with grain. Encourage the dog to move from the flock's head to its tail. When he passes the sheep (as he moves toward the flock's tail), the sheep will move forward.

Call the flock as the dog moves away from the lead sheep. As the dog moves back and forth you will see the flock move toward you. Call to the sheep and step back as they reach your position. Continue moving back as the flock moves forward.

Be sure that the dog moves the entire length of the flock. If he does not, the flock may split into two separate groups, one following the handler and the other remaining stationary, cut off from moving forward by the position of the dog. As you reach the end of the boundary, reverse direction by sliding past the dog on the boundary-line side of the flock. During the reversal give the lead sheep enough time to make its way to the front of the flock.

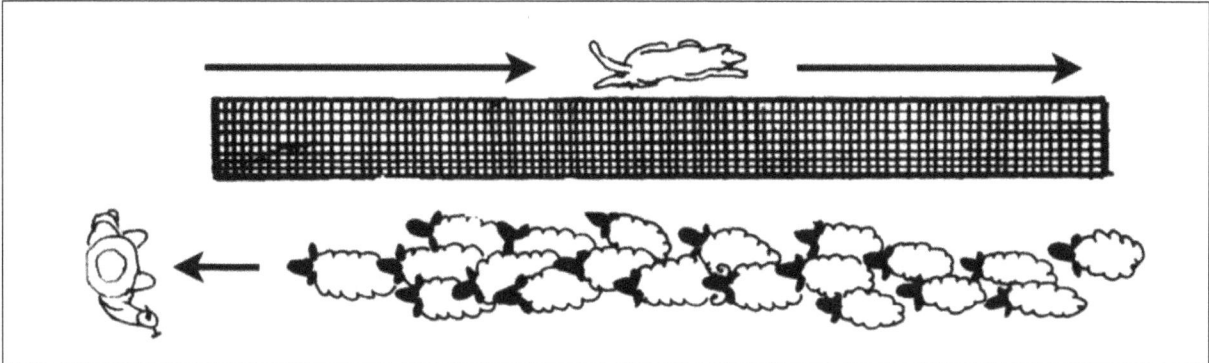

Figure 9.5

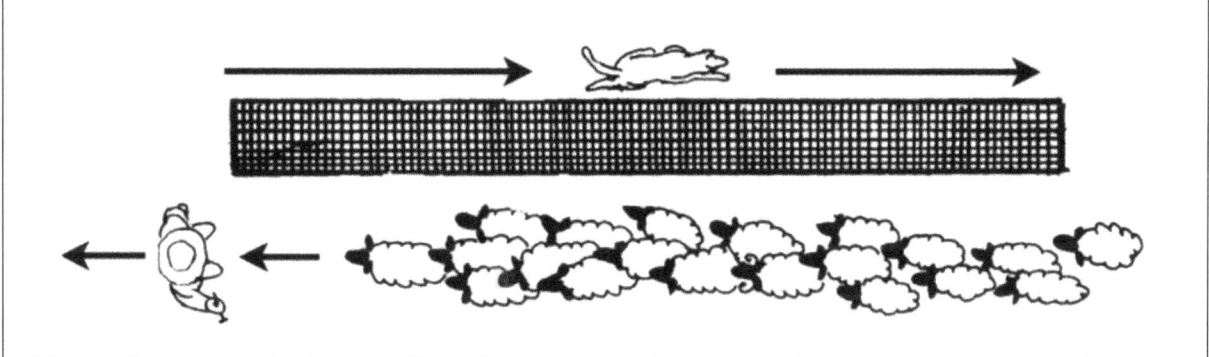

Figure 9.6

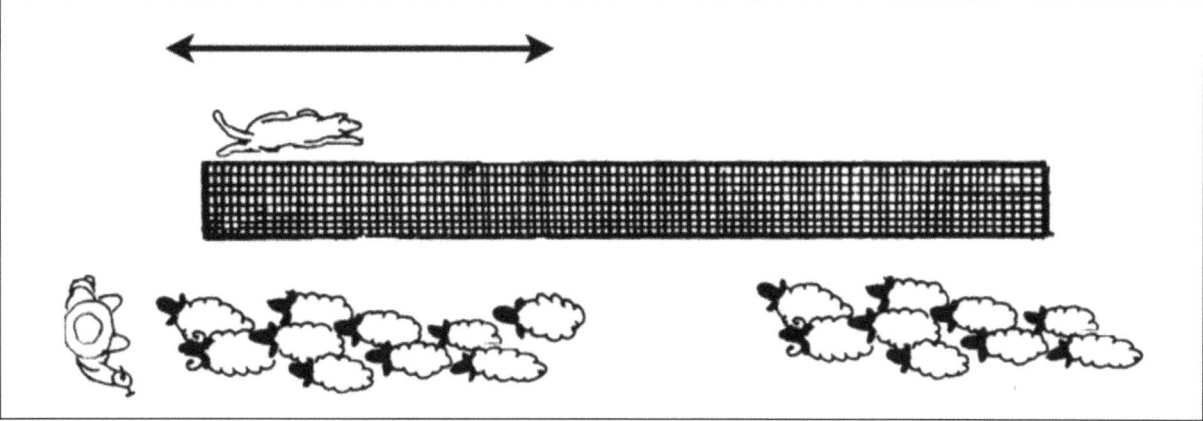

Figure 9.7

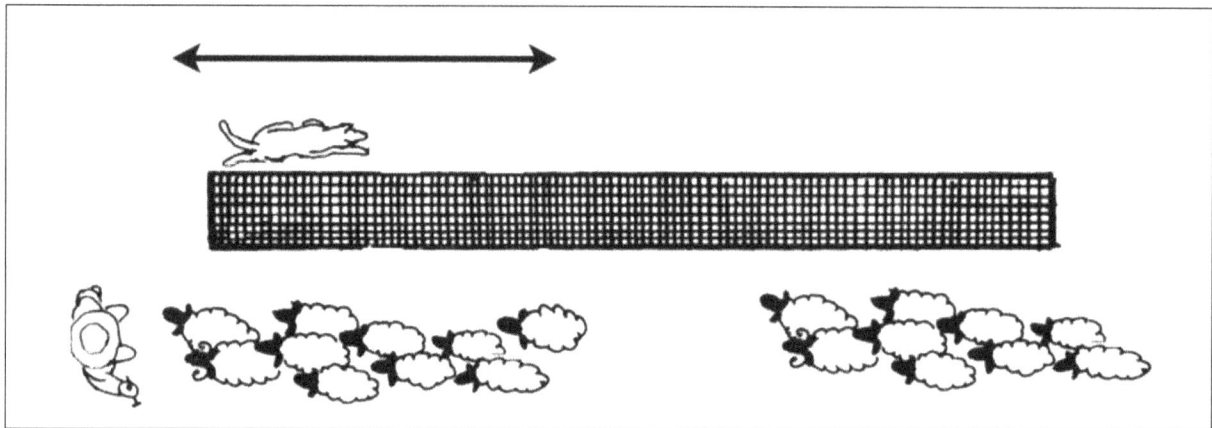

Figure 9.8

When you reach the opposite end, it will become the new head. Encourage the flock to follow you in the new direction. Repeat this exercise at the one hundred foot length until the motion becomes smooth. When both you and your dog feel comfortable with the flowing motion of the flock, increase the boundary length to two hundred feet.

When the flock begins to move with a smooth, flowing motion, you will see it stretch, forming a single column. Your control of the dog at a distance is critical here because he may be one to two hundred feet (or even more) away from you. Continue to practice all your commands, but especially practice the "stand-stay" which is particularly helpful in this exercise.

If the flock splits into two or more groups because the dog does not cover its entire length, stop him just before he attempts to turn toward the stock. Then command him to continue further along the entire length of the flock. This will teach him to traverse the flock's entire length.

Off Lead

Begin each session with on-leash practice and then take the leash off for short periods of time. Increase the length of time the dog is off-lead as he becomes more reliable. As long as he does not get into trouble with his new-found freedom, you may send him away toward the sheep and call him back, but only for short periods of time, approximately five minutes during a training session. Remember to keep the entire session short, no more than twenty minutes in length.

If the dog gets too excited or out of control, put the lead back on him for the rest of the session. At the end of the exercise continue working on the "stand-stay" command with the dog facing the flock.

Chapter Nine: Basic On-Stock Training

Photograph 9.3 Belgian Shepherd grazing 100 plus head along a roadway in Alberta, Canada.

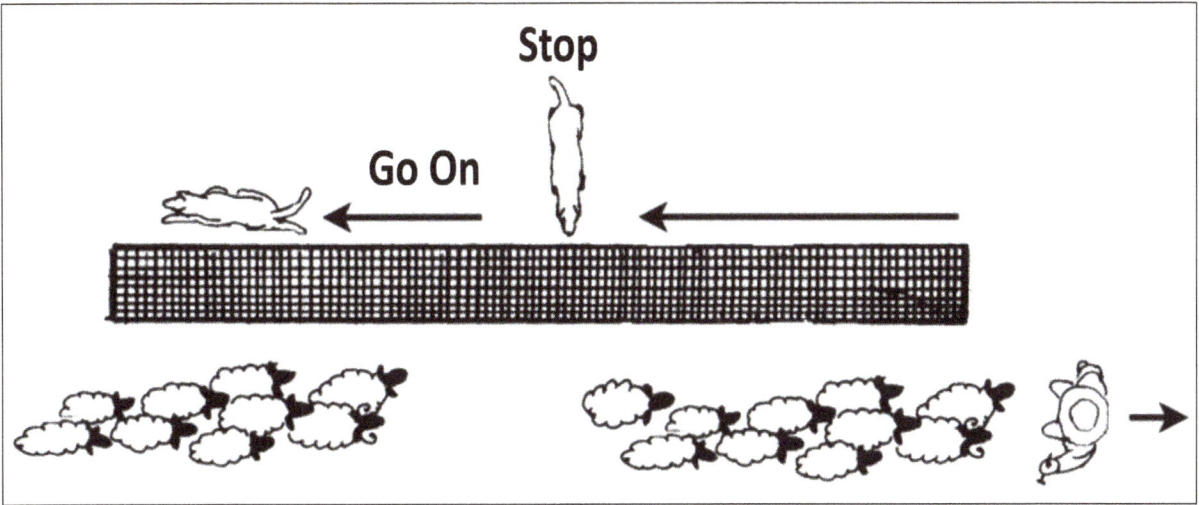

Figure 9.9

Gripping

If a dog does not know how to assert his authority over the sheep, the greedy flock will immediately press forward more and more eagerly, and finally force a weaker dog to stand back, or will even trample him under their feet if he dares to stand fast. Weak dogs are therefore useless for hard work on narrow pastures, and for intensive tending; they can, accordingly, only find employment with the flocks of lambs on large sheep breeding farms. These lambs have their special pastures to mature them quickly. The "lamb dog" is one who must never grip, but only push them with his nose, and he can lie in the shade the whole day long by the side of the equally lazy "lamb boy" (Von Stephanitz 1925:302).

When a dog grips, i.e., catches and holds sheep, it should be similar to a bitch correcting her pup. Allow him to grip to punish the stubborn individual for leaving the flock. This grip is not flesh piercing; rather it is a holding technique. Dogs should grip on the neck, ribs and/or thigh of the sheep depending on the breed of dog. German Shepherds use all three techniques while Briards generally prefer a neck hold. In Europe shepherds sometimes file dogs' canine teeth to reduce the possibility of injury.

Usually it does not matter which of the three kinds of grips a dog uses, so long as he grips in the right way. Good shepherds never allow wrong grips, i.e., those on the ears, throat, breast, shoulder, belly, legs, or tail (Von Stephanitz 1925:304).

To teach a dog to grip properly, place a sheep in a small pen. Have a helper hold the sheep, presenting to the dog the area to be gripped. With your voice encourage the dog to take hold of the area presented. He should only hold the sheep; he must not shake or tear. Correct the dog if he grips in the wrong location or grips too harshly (this is an instance when you will need to use some form of positive punishment). And praise him for making a grip in the correct location with the correct amount of pressure. To determine if there is too much pressure, watch the response of the sheep. Do not allow excess pressure that will injure the stock. To release the dog from the hold call the dog to you or use a release word. In time the dog will take hold of the sheep on your command (see Figure 9.10).

Passive and Active Control of the Flock

Passive control of the flock is the majority of the dog's work. He prevents harmful situations from developing by being in the right place at the right time. Passive control requires no physical contact with flock members. For example, the dog stands in certain locations or patrols a boundary to keep the flock from straying into dangerous areas. Non-passive, or active, control is used only when needed, i.e., to punish sheep who have left the flock.

To begin proofing the guarding or patrolling ability of your dog, allow the flock to stray a short distance over the boundary line. You can accomplish this by placing feed a short distance (2-3 feet) from the boundary line. Encourage the dog to move toward the sheep, thus persuading them to return to the safety of the boundary area. If they do not return, encourage the dog to use his grip as punishment.

By now the dog has learned to passively control the flock by standing in certain locations and by patrolling the boundary line. By trotting back and forth along the boundary line the dog approaches the stock from the side, not head on. Over time, seeing the dog approach will condition the flock to return to the grazing area. If the dog moves in between the boundary line and the escaped sheep, he will cause the sheep to scatter farther away from the grazing area. The individuals that have left the security of the main flock will now feel cut off from it and will panic.

If the dog executes the approach correctly, the sheep will have a direct route back to the main flock. The flock will not be cut off from the boundary line, thus allowing the sheep to move back into the boundary area. The dog must move toward the flock with authority.

Figure 9.10 "Grips"

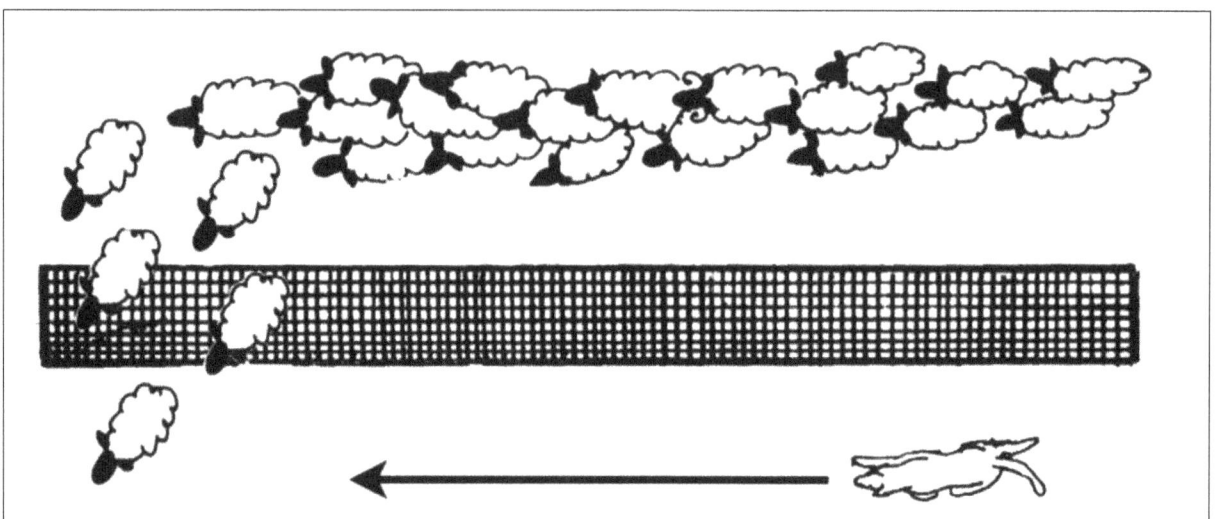

Figure 9.11

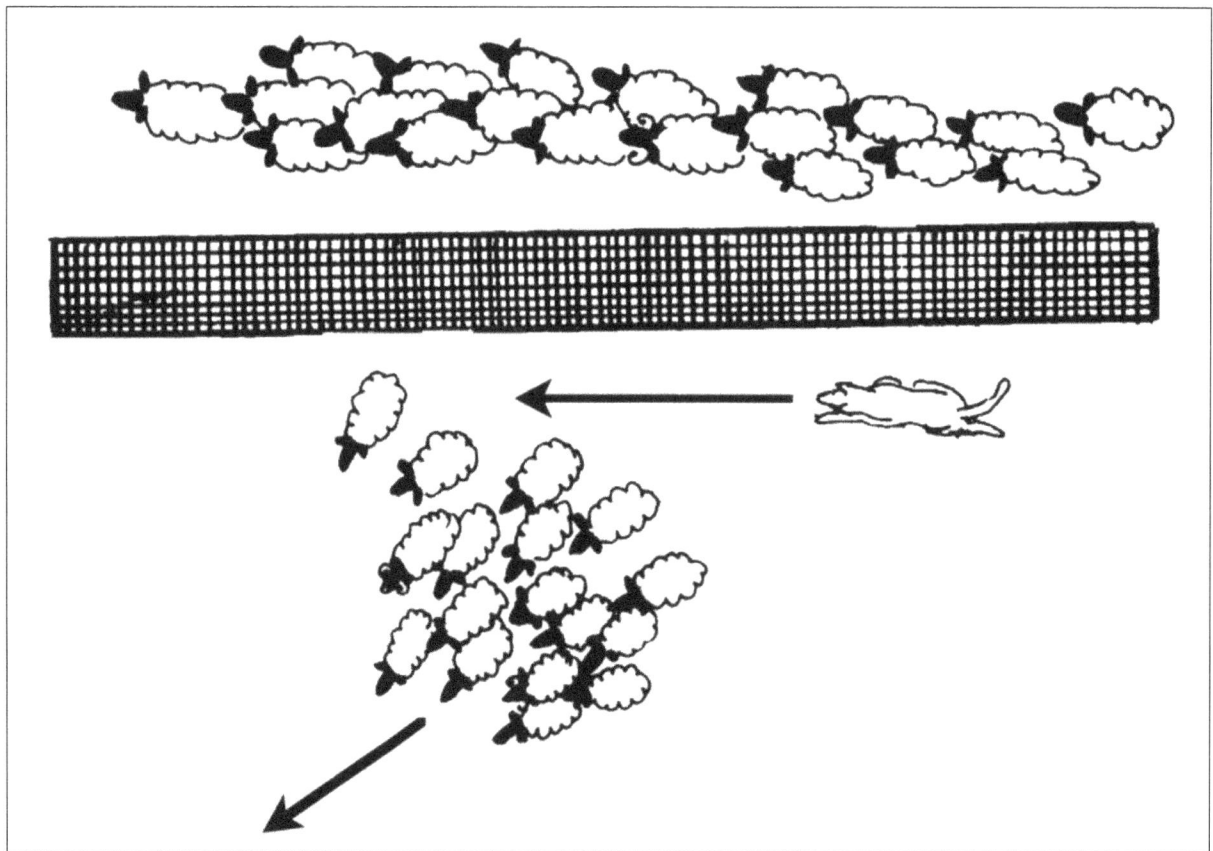

Figure 9.12

To insure that the dog learns the correct approach, allow the flock to spread over the boundary line. Give the dog an opportunity to perform the task correctly. The sheep outside the grazing area will move back into the grazing area with ease, creating a positive experience for the dog (see Figure 9.13).

As the dog gains experience, increase the distance the sheep are allowed to stray beyond the grazing area. When the dog approaches the escaped sheep the handler should move to a position behind the dog, holding on to a long line attached to his collar. At first you must guide your dog to the correct position to push the flock back into the boundary area.

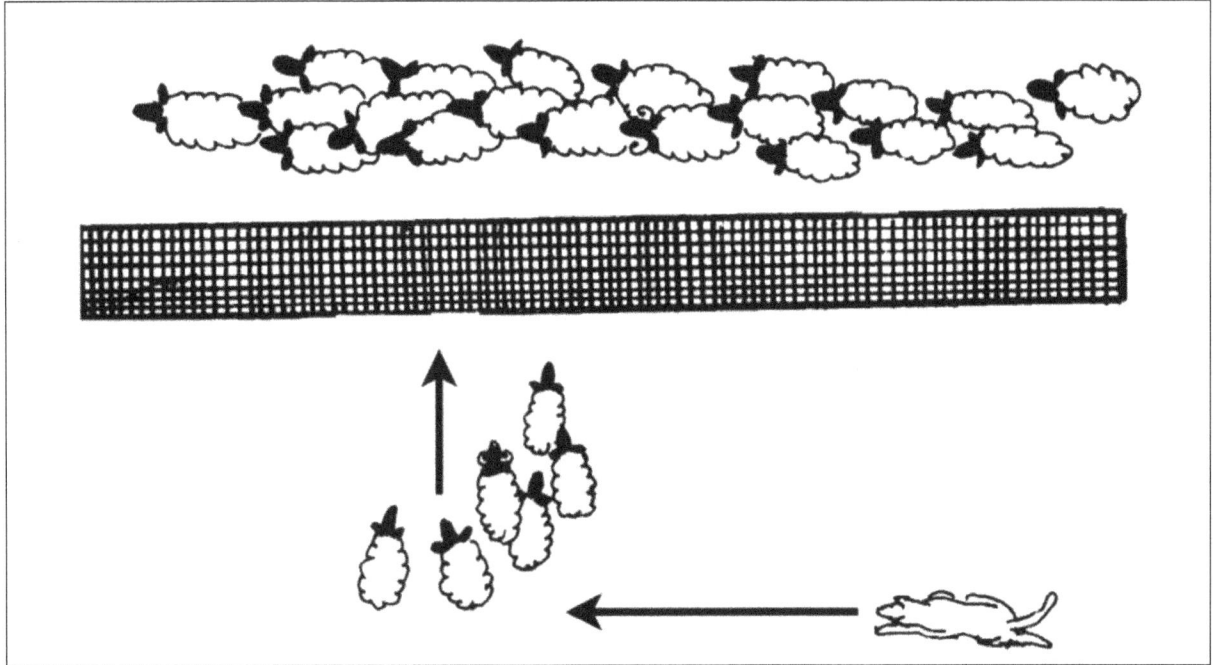

Figure 9.13

Whether or not the dog is allowed to grip is your decision. If you do allow gripping, teach the dog to do so only on command, and use it only as a last resort when passive prevention has failed. There are times when you need to remind the flock that active control exists as a way of controlling them.

Usually the flock will not stray too far from the boundary area because sheep move slowly as they graze. Consequently, it is not difficult for the keen tending dog to keep his charges from straying too far as he patrols the perimeters of the grazing area (see Photograph 9.4).

The Corner

Work at the corner lays the foundation for pen, bridge and road work discussed in the next chapter. To begin, stand the dog inside the pen near its mouth. Lead the flock out of the pen and release the dog. Immediately step to the side of the flock and call the dog straight to you. You will need a second handler to control the flock as you step aside.

It is important for the dog to come straight to the handler. If he does not, the dog will run into the flock. When the dog reaches you, have him walk to the front of the flock with you. The second handler should now leave the flock in your control. The flock should already trust the handler; now they must learn to have confidence in the dog. Consequently, he must not try to punish the flock during this exercise.

Working the corner of a grazing area requires the dog to be some distance from the handler. To avoid initial problems, first return the sheep to the pen. Then, begin working on one side of the pen, practicing all of the previously learned commands. That the dog has learned to turn toward the stock in previous exercises will greatly aid in teaching the corner technique. As you flow back and forth patrolling the inside of the pen, you will begin to notice that you pull the dog along with you as if he were on an imaginary lead. When the dog contacts the corner of the pen he will probably notice he can continue along the fence to the other side of the pen.

If he does not do this on his own, you can easily show him how to negotiate the corner. Position yourself at corner "A" (see Figure 9.15). The dog trots back and forth between corners "A" and "B". Ask the dog to "stand-stay" at corner "B". This will allow you time to position yourself at corner "C." Call the dog toward you. This will pull the dog around the corner. Repeat this from both directions until the dog understands how to negotiate corners. When you can place the dog at corner "A" and call him to corner "C," you are ready to move to the next phase of the exercise.

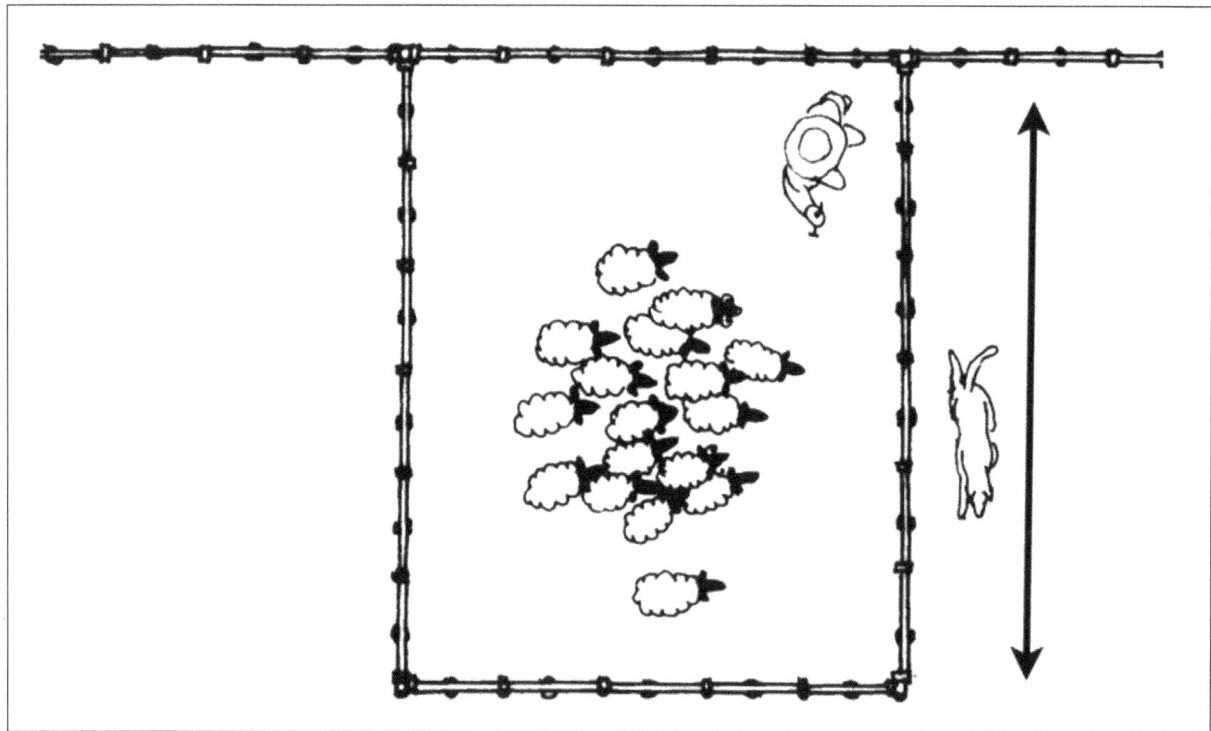

Figure 9.14

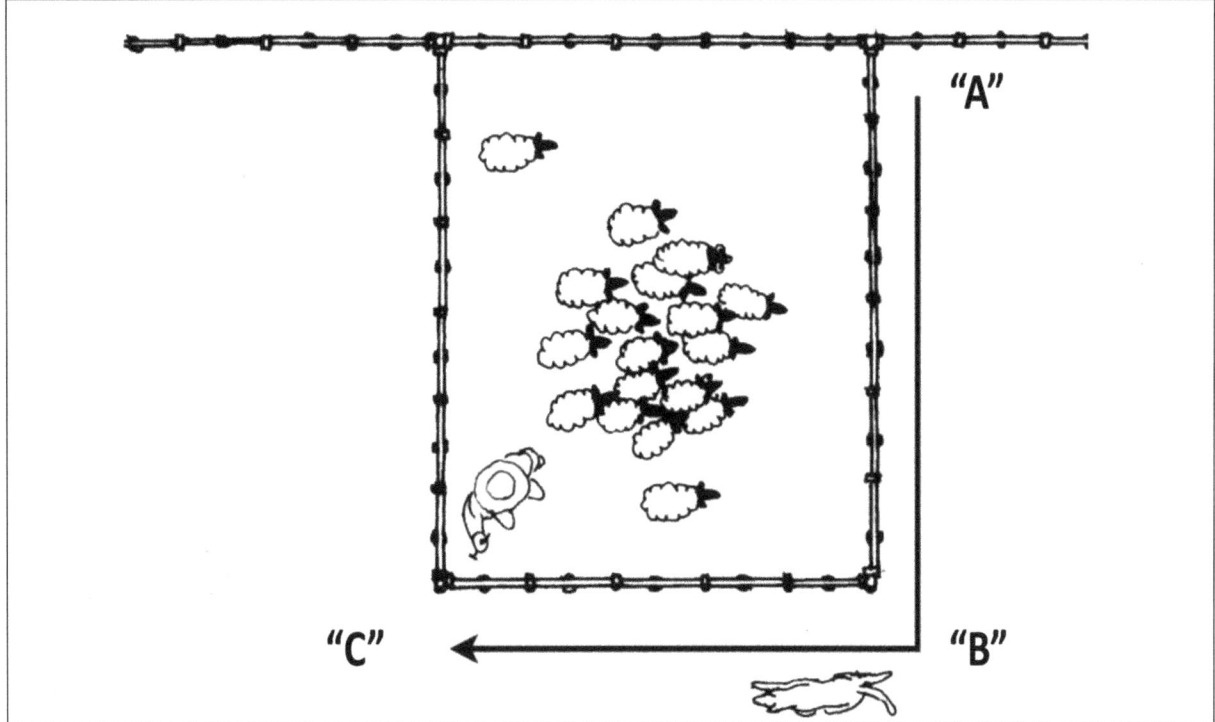

Figure 9.15

Two Sided Graze

To teach the two sided graze we use a new configuration of boundary lines in the shape of an "L." We place the "L" against a corner of the fence to form a square, which the dog will patrol. I suggest each side be fifty feet in length.

To begin this exercise, lead the flock into the boundary area. You should position yourself at the corner of the two boundary lines. Encourage the dog to patrol one of those sides. During this time practice the "come here," "go on," and "stand-stay" commands.

Once the dog is performing consistently, encourage him to come to you and pause. Then encourage him to move around you to begin patrolling the other side. You may wonder why you must stop the dog before he turns the corner when teaching him to turn the corner is the purpose of the exercise. The answer is simple. By making your dog anticipate the stop you teach him to come into the corner at a slower pace. Otherwise he might increase his pace due to the increased distance he must now patrol.

As the dog works one side and then the other, he will begin to work the corner without realizing he has come around it. Now, slowly decrease the number of times the dog patrols one side before asking him to work the other. Do this until he moves from one side of the "L" to the other in a single fluid motion. You may find that the dog cuts the corner of the boundary, but by positioning yourself one or two feet from the corner, outside the boundary area, you can correct this. The dog will now have to go around you as he approaches the corner. If he does not go around the corner, you will be in position to help him do so. Using your voice, coax your dog around you. Keep repeating the corner exercise until working the corner correctly becomes second nature to the dog.

The Pasture Graze

In general, if the flock is quietly feeding or the sheep are moving in order, the dog should not disturb them. Instead he should remain quietly in his place patrolling the boundary. Only if he sees the sheep advancing toward crops, should he ward them off (Von Stephanitz 1925:301). The pasture boundary exercise combines what you and your dog have learned so far. Because the boundary has three sides, now the sheep can escape in three different directions. Consequently, the handler and the dog must be even more observant than before.

By this time the sheep should have learned that it is better to be inside the boundary area than caught outside and punished by the dog for leaving the assigned area. If the sheep do leave the area the dog should correct them. Usually, however, the sheep will return to the safety of the boundary area when the dog closes to about twenty feet of them.

You may notice that your dog keeps to one or two sides of the boundary area. He should. For he is feeling pressure from the sheep who want to cross these sides of the boundary area to get to "greener pastures." As long as the other members of the flock do not leave the boundary

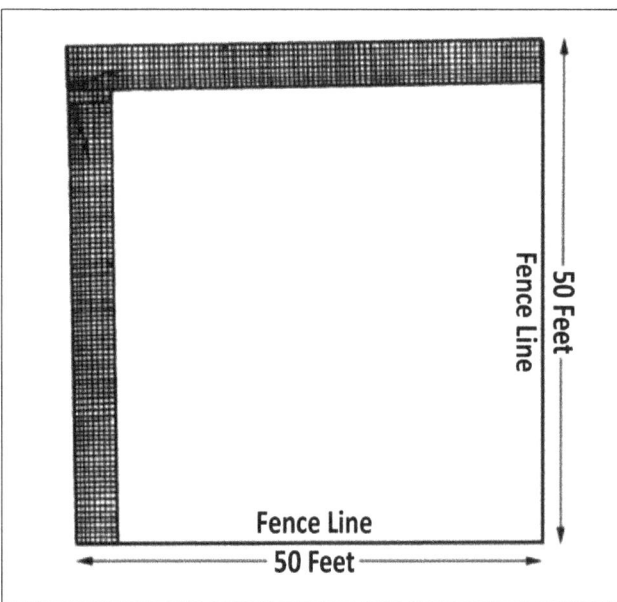

Figure 9.16

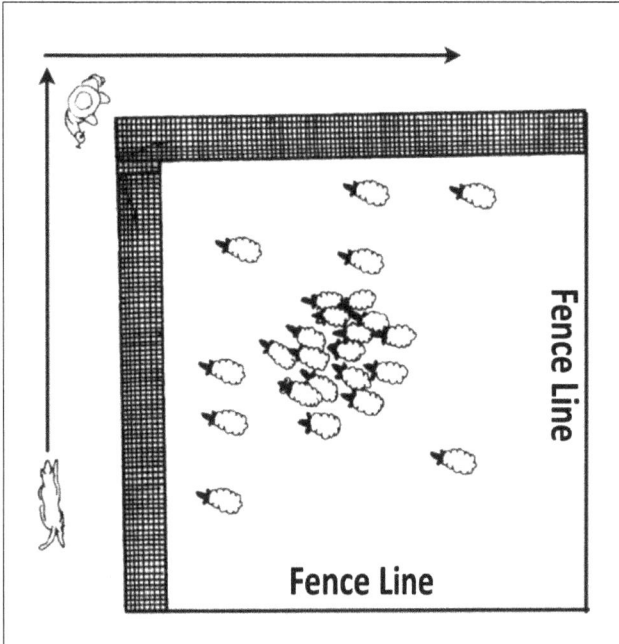

Figure 9.17

Photograph 9.4 While patrolling a boundary in Germany, this Briard became aware that some of the flock had strayed. She quickly returned them to their assigned grazing area.

area on either of the un-patrolled sides, the dog may continue patrolling the side of his choice.

You should think of the boundary area as two "L"s, not a three sided area. Position yourself at one of the corners and begin the exercise with your dog patrolling one of the "L"s (i.e., two sides of the area connected by a corner). After he settles and is working smoothly, move to the other corner of the boundary area and repeat the exercise with the dog patrolling the other "L." When you begin working a four sided grazing area, you will need to use two dogs, each patrolling his own "L" (see Figures 9.19 & 9.20). You must discourage your dog from mindlessly circling the grazing area. This can happen if you omit steps in his training and instead skip to the pasture boundary area exercise.

In actual working situations where shepherds are with their flocks for entire days, practice is easy and relaxed, for the time intervals between sheep straying from boundary areas to graze are often intermittent. It would be best to wait for training opportunities to arise naturally in the course of the working day, but most of us cannot wait for such opportunities. We can, however, compensate by placing extra feed around the boundary area at different locations to tempt the flock to leave the area.

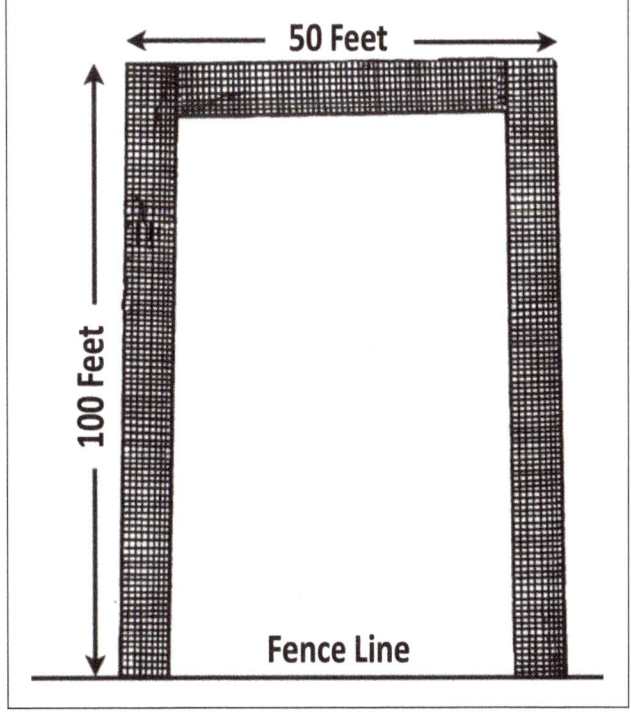

Figure 9.18

If the sheep are in the center of the boundary area grazing peacefully the dog does not need to continually patrol it. In such cases he should station himself near the side of the boundary line(s) closest to the sheep, patrolling as necessary. Remember, dogs must save their energy if they are to guide their flocks for the entire day. It is all right for your dog to be at ease as long as he stays alert, prepared to react as the situation develops.

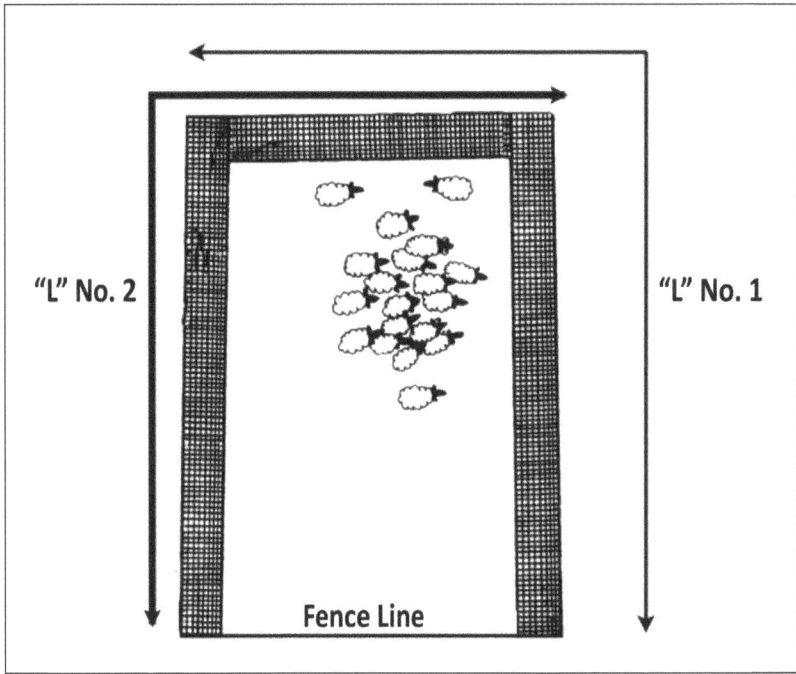

Figure 9.19

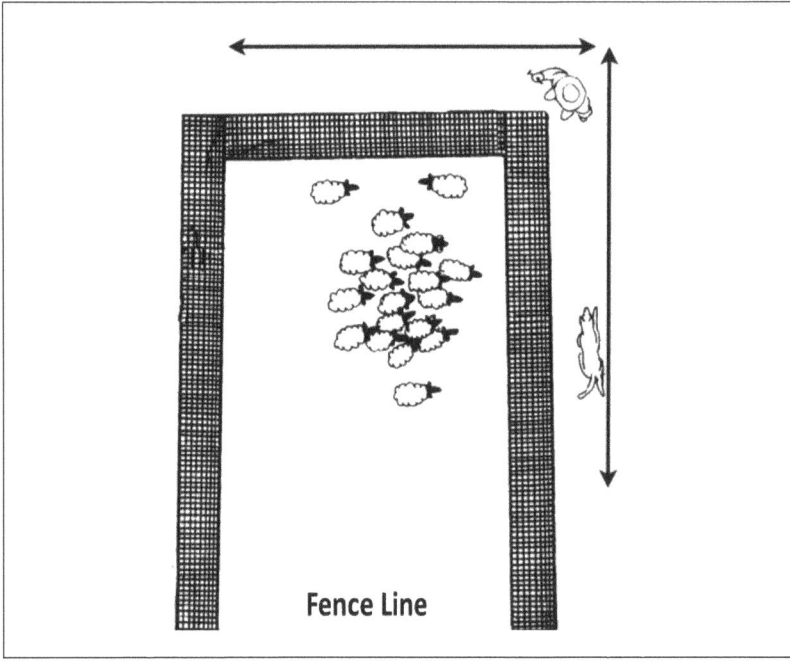

Figure 9.20

The Four Sided Grazing Area

There are two types of four sided grazing areas: wide and narrow. The wide grazing area is usually rectangular in form and approximately 25 by 35 yards. Narrow grazes are smaller, approximately 10 by 100 yards. In four sided grazing areas allow the sheep to spread out in a more relaxed eating formation. The dog will have a large area to oversee but because the flock is likely to be content, the dog can relax.

When you first enter the grazing area, only step in a few feet. At first the flock will stop but in time the sheep will notice that the area you are in has choice grass and they will begin to graze forward. The sheep don't know where the boundaries of the area are. Consequently, you and your dog must show them. While the sheep are grazing, stay as stationary as possible, either alongside the area or at a corner. As you discovered when entering the area, the sheep will keep their eye on you, taking their cue from you, their leader. When you move, they move; when you stop, they stop. If the flock moves forward too rapidly, however, they will waste good forage by trampling it. They will act like children in a candy shop, sampling everything and wasting most of it. To minimize this send your dog ahead of the flock in a "wide arc,"[33] stopping him directly in front of the flock's forward motion (see Figure 9.21).

Photograph 9.5 **German Shepherd Dog working with an Old Fashion German Shepherd, each patrolling its own "L".**

---

33 The first time I heard "weiten bogen" (wide arc) in conjunction with training tending dogs, I pictured a dog moving out in a curved path similar to the outrun of a Border Collie. But the word "arc" is misleading here. In this context it does not mean a curved path (a segment of a circle). On the contrary, the dog moves in straight lines and right angles. The arc is squared off, as it were. Similarly the dog's placement before the flock looks like an outrun, lift and fetch. But as we have seen, the dog does not move in a pear shaped outrun and does not lift the sheep and fetch them to the handler. On the contrary, he moves along the boundaries of the grazing area, and when he reaches the top he stops and turns at a right angle before moving toward the sheep (see Figure 9.21).

Placement Before the Flock

You will now stop the flock's forward movement by commanding your dog to leave the boundary line and slowly walk toward the balance point (head) of the flock. Not only will this stop the flock's movement, but it will make the grazing area smaller as well. The command to walk in toward the flock will be "walk on."

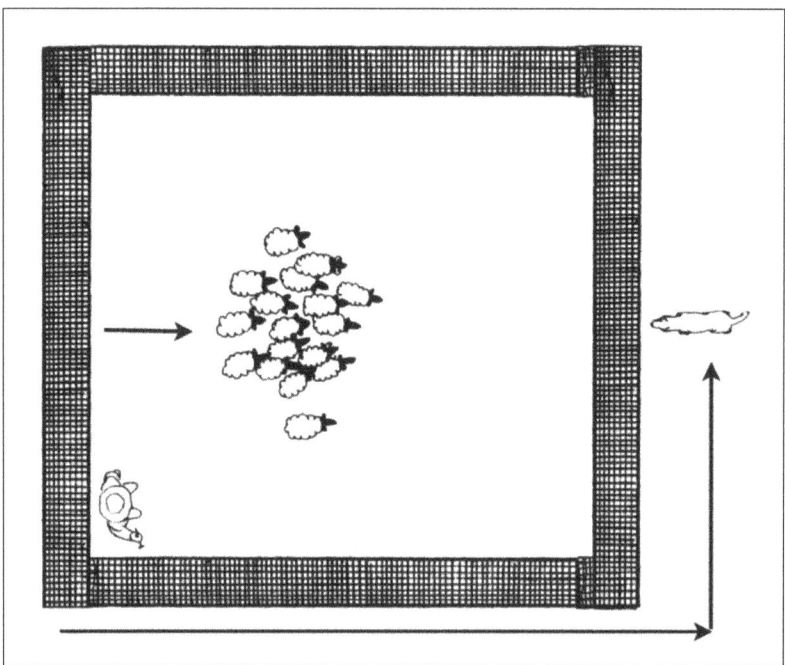

Figure 9.21

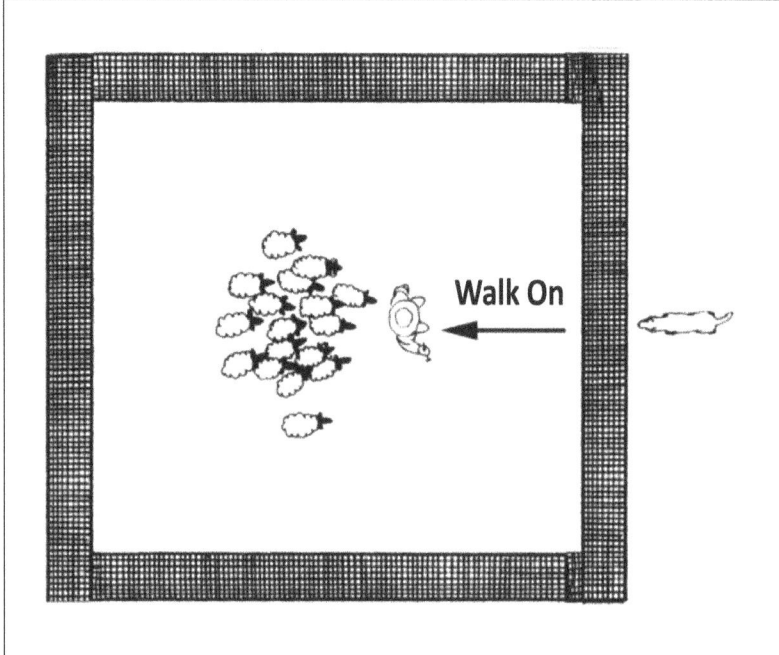

Figure 9.22

In most cases the dog will not want to cross into the grazing area because earlier you taught him not to do so. Consequently, you may need to coax him into entering the graze. To do this, position yourself in front of the flock opposite the dog. Then get down on one knee and make you and your voice as inviting as possible, calling the dog with the "walk on" command. You may have to use his name as well as the command. As the dog crosses the line, praise him for a job well done.

After the dog enters the grazing area he must concentrate on his charges. His eyes should be on the flock, not on you. Step away from your position in front of the flock and gradually increase your distance until you are on the edge of the grazing area once again.

Once the flock stops its forward motion, the dog has completed his task. Do not use the "out" command at this time. Instead, escort him back to the boundary from which he came. He may take "out" as a reprimand and feel he was wrong in coming into the grazing area. You may use it later, however, when your dog is secure in entering the pasture on command. Be sure the dog returns along the same path to the boundary line.

Practice starting and stopping your dog along the route toward the flock. He must not rush forward, scattering or turning the stock around in the process and he must not turn this into a lift and fetch exercise where he lifts the flock and fetches it to you. This is a passive exercise—an obedience workout for your dog.

Bring the dog as close to the sheep as possible without disrupting their foraging. The sheep at the edge of the flock may pick up their heads to turn away from the dog. But they should soon return to grazing, though not in the area cut off by the dog. Although the flock will spread out while foraging, they should graze the area as efficiently as possible.

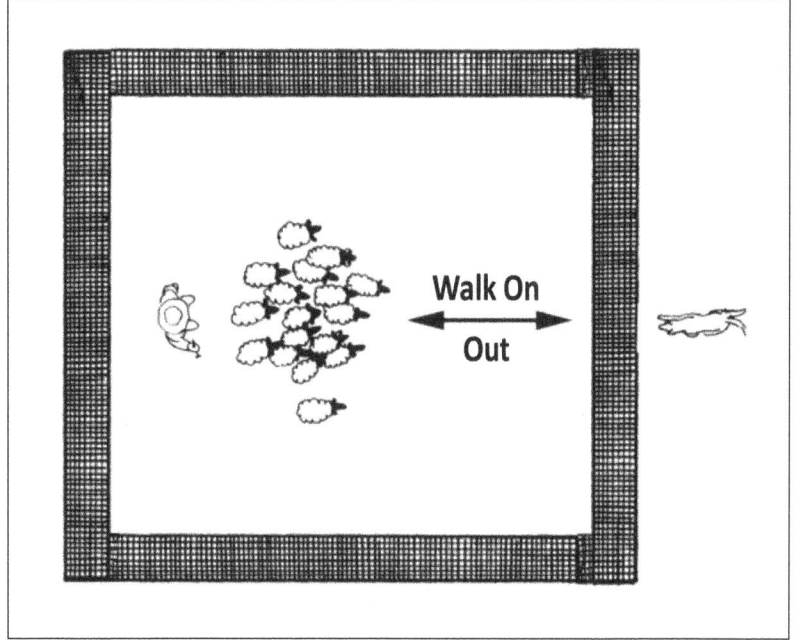

Figure 9.23

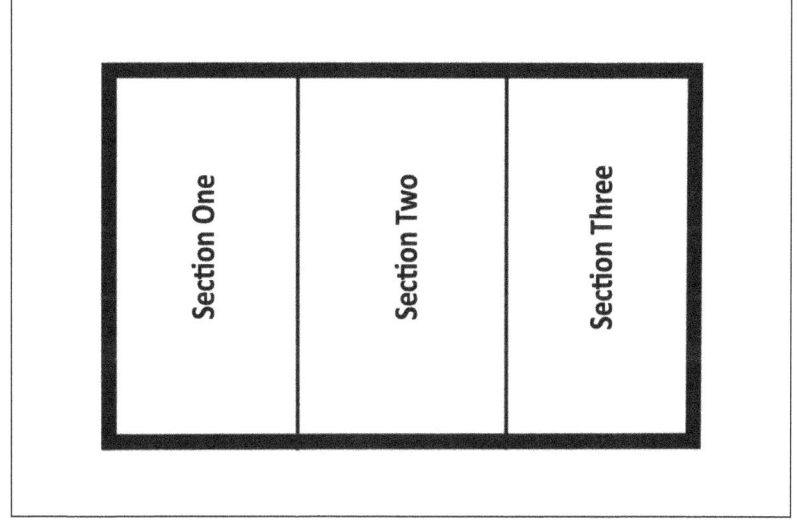

Figure 9.24

In your mind divide the area into three sections. Systematically graze the first, then middle, and finally, the last section. If you use this technique the pasture will last longer and the flock will have more grass to eat.

Narrow Graze

When warding off in front of the crop, the dog must never make unnecessary exertions; he walks up and down the furrow before the pasturing flock when the occasion demands it, this marching is called, "sentry-go," and stands from time to time to convince himself that all is in order behind his back. He will drive off single pilferers by a short run, and the obstinate ones are punished by a short firm grip (Von Stephanitz 1925:302).

The narrow grazing area is used for intense rotational strip grazing. Here the objective is to move stock forward while grazing. In narrow grazing areas the stock are not allowed to spread out while foraging but enter one end of the pasture and mow the grass as they move forward, much like a mechanical mowing machine. The dog's task is to keep the flock in formation while the shepherd's job is to control the pace of the forward motion. In other words, the shepherd controls the flock's direction and speed, while the dog controls the flock's length and width.

Photograph 9.6 German Shepherd Dog quietly standing in front of its charges. Notice that the flock is still eating and have not turned away from the dog.

Photograph 9.7 A German Shepherd Dog taking time out to oversee its charges.

Narrow grazes, as the name implies, are long and narrow, approximately twenty to thirty feet wide. Like the road ways they resemble, the long boundaries (similar to the sides of a road) are the most critical ones. The front and rear are less important.

Upon entering a narrow graze, begin slowly walking along one side, staying even with the last third of the flock. Immediately send the dog in front of you and around the front of the sheep so he can effectively oversee the opposite side of the flock. You take responsibility for one side while the dog assumes responsibility for the other. You can use the spade end of your staff to help control the flock on your side of the graze by tossing dirt in front of the sheep to keep them contained.

Traditionally, shepherds used two dogs in narrow grazes: one to help with their side of the flock, and the other to patrol the opposite side. When using one or two dogs, the main dog works the side opposite the handler. It takes more skill to balance the flock within the grazing area with one dog than it does with two.

You should be careful to practice changing sides with your dog (see learning the "switch" command in the next chapter). There will be times when the dog is needed on the opposite side of the flock, especially if you are working with one dog. The dog must work on the side of the flock that poses the greatest danger to the flock or crops.

The dog should not cut in too close to the front of the flock but move in a "wide arc" (see footnote on page 169). Otherwise he will stop the forward movement and interrupt the sheep's grazing (see Photograph 9.9). Once the flock reaches the end of the grazing area, lead them to another grazing area and repeat the exercise. Balance wide graze with narrow graze work.

The flock can run out of practice very quickly if conditioned only to wide grazing.

Your position in the narrow graze is very important; you should stay at the edge of the grazing area. Walking around in the middle of the grazing area will confuse the flock about where they should go and in time they will not respond quickly either to your calls or to your forward movement. Remember, you and your dog are a team; he watches one side while you control the other. If you move to the front of the flock, it alerts both the dog and flock that it is time to increase the pace and move forward.

During trial competition the dog is required to work the side of the flock opposite the shepherd in both the narrow and wide grazes. In both situations the dog should immediately go to the opposite side of the flock upon entering the graze. Too much pressure from the dog can push the flock over the edge of the boundary, causing a zigzag movement. When two dogs work together they press the sheep between them, thus keeping the flock from zigzagging. A single dog, however, must work more quietly and further away from the flock to maintain balance and avoid zigzagging.

Photograph 9.8 German Shepherd Dog patrolling a narrow graze during a competition.

Chapter Nine: Basic On-Stock Training

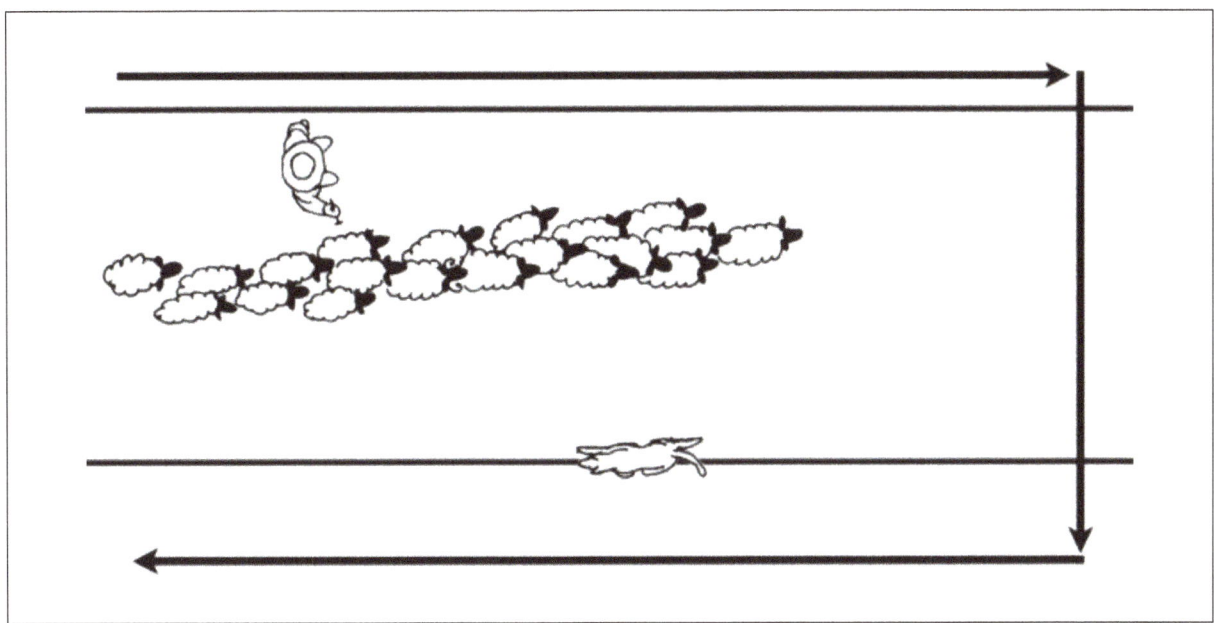

Figure 9.25

Photograph 9.9 Two shepherds changing sides in front of a grazing flock. The dogs changed sides too close to the flock causing them to stop grazing, thus inhibiting their forward motion.

Hear My Voice: An Old World Approach to Herding

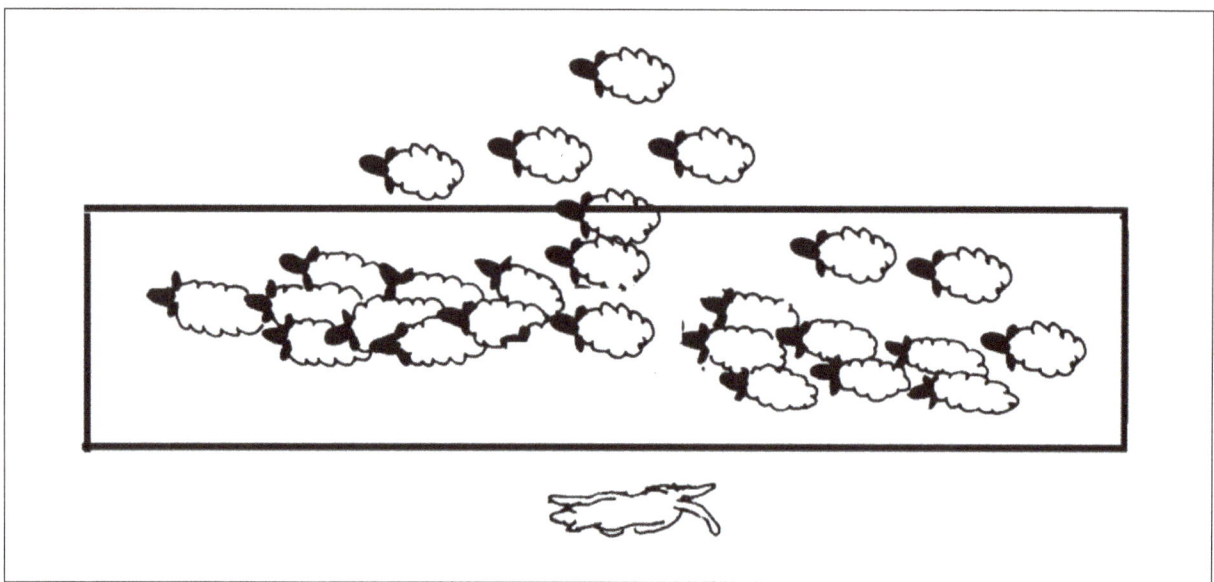

Figure 9.26 Too much pressure from the dog causes a zigzagging movement of the flock. This also causes the flock to move out of the grazing area.

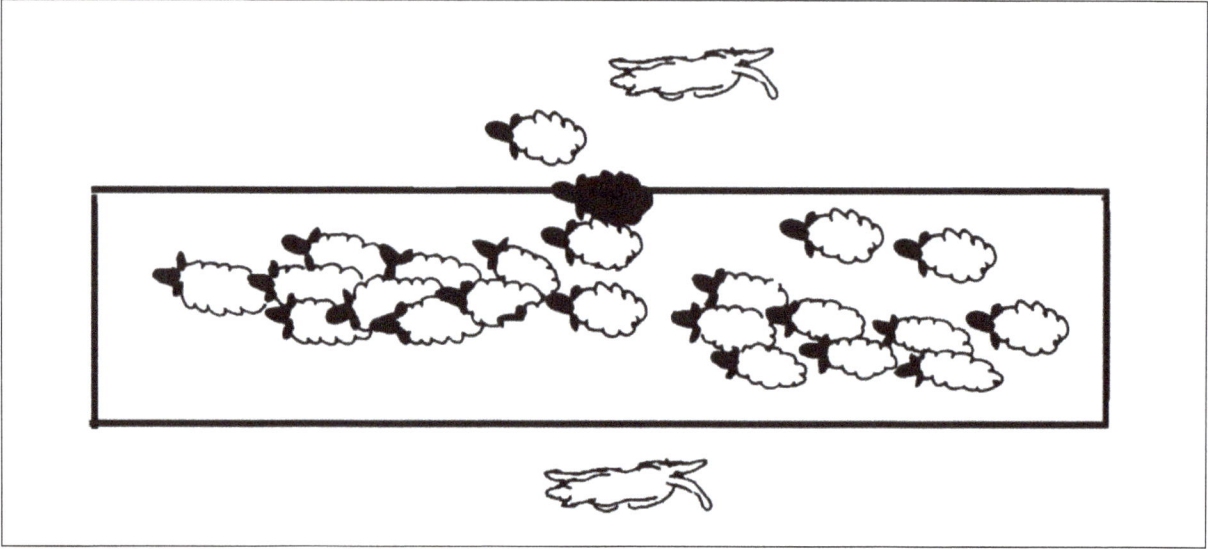

Figure 9.27 Two dogs can easily balance a flock within a grazing area. If a sheep leaves its assigned area the dog on that side can quickly return it back to the flock while the other dog controls the rest of the sheep.

Ten

Advanced On-Stock Training

Now that your dog has learned the basics of tending work, it is time to begin more advanced training. Start with changing sides and going to an object. These are important skills for they underlie turning corners, traffic work and work at the pen, the three advanced, tending exercises that your dog must master.

Changing Sides

I cannot over-emphasize that the dog must always change sides in front of the flock and never at the rear, or by going through the flock on the road (Von Stephanitz 1925: 299).

Photograph 10.1 Shepherd changing sides in front of the handler notice that the lead sheep are not stressed by this.

Roadways and narrow grazes present special problems to shepherds. Often they must move their dogs from one side of the flock to the other to protect crops and/or livestock from harm. I usually teach the "change of sides" exercise on the road because roads have added difficulties. Still, the principles are the same and once a dog has mastered changing sides on a roadway, he can easily accomplish the task in a narrow graze. In the pages that follow I discuss "changing sides in both locations."

Herding dogs that drive and/or fetch sheep naturally move from one side of the flock to the other by passing behind it. But this causes the flock to move faster, which in turn causes the sheep to walk shoulder to shoulder. Consequently, they fan out along the width of the road, increasing the likelihood they will leave the roadway altogether. Moreover, when working with two dogs, you must watch them carefully, for they may fight or play as they pass each other changing sides. If the flock is large (300-1000 sheep) you cannot oversee your dogs if they change sides at the flock's tail (see Figure 10.1). To avoid both these problems, the dog(s) should change sides at the head of the flock, crossing in front of the shepherd who is leading the sheep.

Begin by practicing without livestock, walking your dog both in a furrow and along a road side. When you arrive at the place where you want him to change sides, stop your dog and turn at a right angle toward the other side. Escort him to that side, naming the action "switch." Once you are on the other side continue walking. After enough practice (it shouldn't take longer than an hour in total) the dog will move to the other side of the road or narrow graze when he hears the "switch" command.

When your dog has mastered this command it is time to introduce the sheep. Begin this part of the exercise while working a three-sided boundary area. Remember,

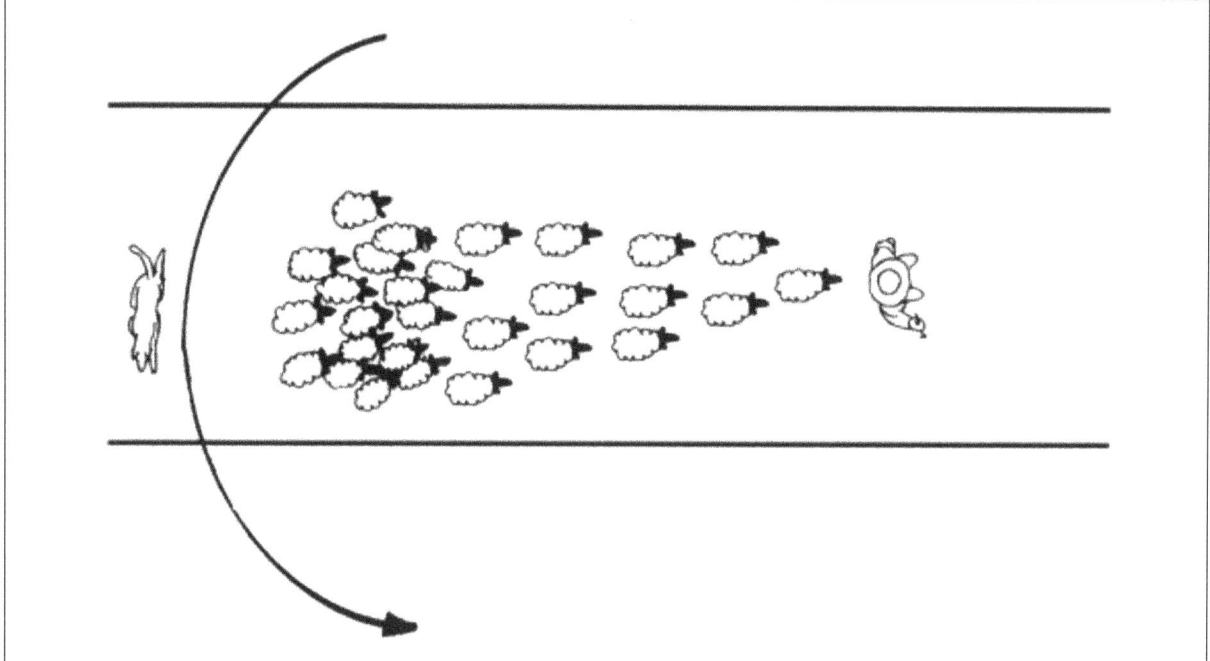

Figure 10.1

for the reasons discussed above, your dog should accomplish the change of sides while in front of the flock.

To begin the exercise, sprinkle corn or other grain along the center of the grazing area. This will hold the sheep in place while you work the dog. Position yourself next to one of the long boundary lines and allow the dog to work that side. Move to the front of the flock, call the dog forward, stand him, and then give the "switch" command. The dog should now cross to the other side of the grazing area and begin patrolling the new side.

In time you can drop the "stand" command before asking the dog to cross to the other side (see Figure 10.2). When the dog changes sides smoothly with only the "switch" command, stand farther back along the last third of the flock. Ask the dog to change sides from there (see narrow graze exercise).

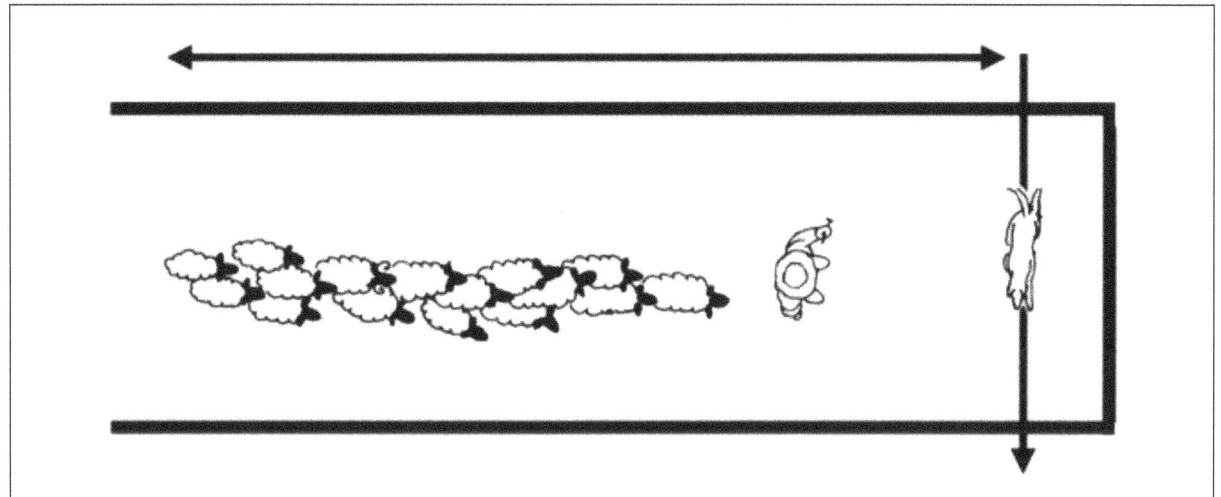

Figure 10.2

Go To an Object

There are two parts to this exercise: the first teaches the dog to go to an object, while the second teaches him to stand correctly in position at that object. The dog must stand where he can see the entire flock and thus must be able to look both to the left and right. You can use this exercise at a pen-gate, bridge, sharp corner or parked car.

To start the exercise, place the dog on a six-foot lead. Walk with him toward the object you want him to go to. In this case, let's make it a panel. As you reach the panel, pat its side and say "pick." Repeat this exercise every chance you get. Make a game of it. The dog will soon learn that the word means for him to go to the object you have chosen.

Photograph 10.2 **This dog was sent to a tree at the corner of a roadway.**

Once the dog willingly goes to the chosen object on command, he must learn to stand in position at this object. The dog should stand parallel to it; be sure his chest does not protrude past its plane.

After you send the dog to an object require him to stand guard with the "stand-stay" command. He must stop in his tracks, oriented in the same direction that he was moving and positioned facing the column of sheep. Practicing this exercise at a panel will help key the dog to stand in the correct position. In any case, you will send your dog to a panel at least half of the time he performs this maneuver. Most of the other times, you will send him to a corner location. Be sure to practice working the dog both on your right and on your left.

Left and Right Turns

The most important command in this exercise is the "stand-stay." Place the dog in a "pick" position at the corner of a two-sided boundary area. In this position the dog can see the entire flock, front and rear (see Figure 10.3). Now, lead the flock around the dog that is positioned at the corner. Be careful not to make your turn too tight. If you do, the rest of the sheep will see those ahead of them going in their new direction. Consequently, these sheep, i.e., those who have not yet made the turn, will cut the corner short, failing to go around the position correctly. On a roadway this could have disastrous consequences (See Figure 10.4).

Begin the next phase of the exercise by leading the flock along one side of the boundary area. As you approach the corner, call the dog to it with the "pick" command. Then stand him. With a flock new to tending you may need to stop the dog short of the corner. Thus, the flock will start its turn before you have set the dog in his correct position. This allows the flock to approach the corner without the sheep feeling they are not allowed to pass. As the flock gains confidence in you and your dog, you will be able to position the dog in the proper place at the corner before pivoting the flock around him. Continue to call to the flock as you pivot the corner.

When the front of the flock makes the turn, bring the dog to the correct position. The dog should hold this position until the entire flock has passed.

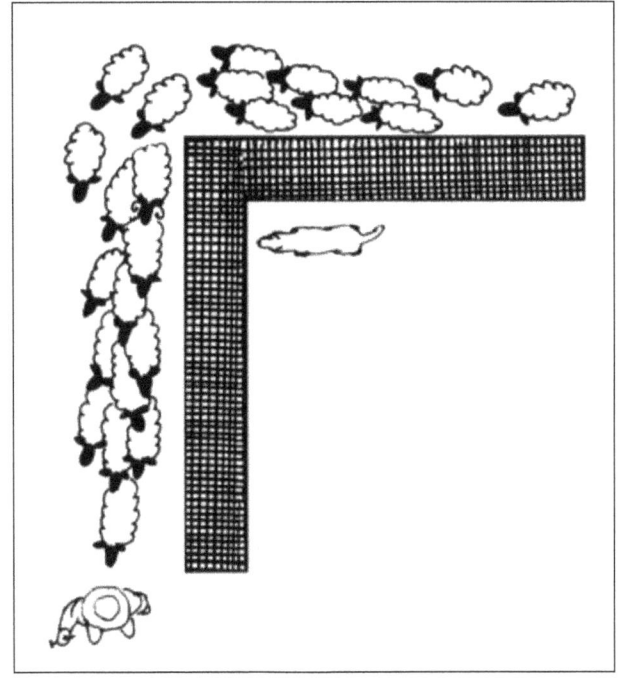

Figure 10.3

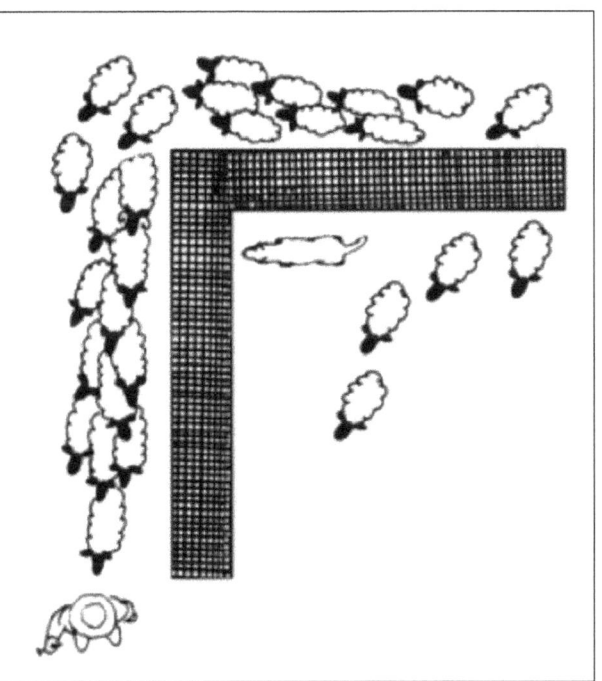

Figure 10.4

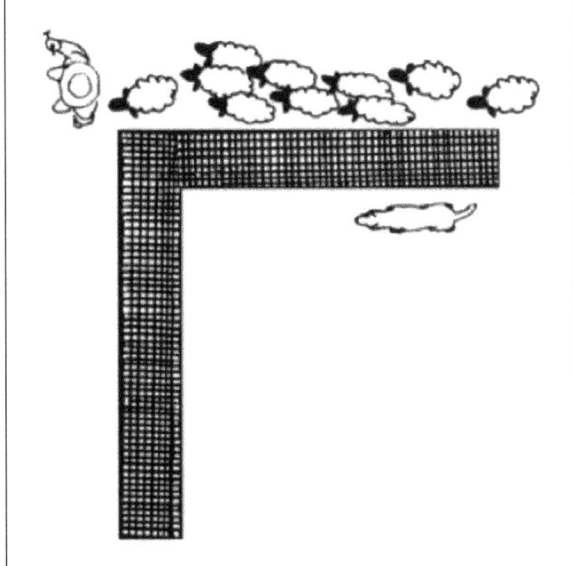

Figure 10.5

Photograph 10.3 A dog is used as a pick as a flock exits a wide graze.

This exercise demands timing and patience from both you and your dog. You must work as a team. After you become proficient, the sheep will pivot smoothly around the stationary dog as they turn the corner.

Road Work

They know how to 'balance' the sides of the flock and they will also know how to prevent the sheep from leaving the road to graze on the shoulder or in the neighboring fields (McLeroth 1982:175).

Now that you and your dog have mastered the "switch" and "pick" commands you can commence working the flock along a road. In some ways moving sheep along a road is similar to moving them through a narrow graze pasture. In other ways it is significantly different, for sheep move much faster along roads than through pastures. This is because their forward motion is not slowed by grazing, even though sheep might stop periodically for a bite along a roadway's edge. Moreover, the narrow width of some roads may further increase the flocks pace, making it difficult for novice shepherds to control both their dog and the movement of the sheep at the same time.

The best way to deal with this problem is to ask a second handler to lead the flock while you work your dog along the flock's sides. As your dog becomes better trained, and as you become more proficient in your shepherding skills, you will learn to concentrate on both your dog and the flock at the same time. You can then dispense with the help of the second handler.

Chapter Ten: Advanced On-Stock Training

Photograph 10.4 Two German Shepherds balancing a large flock on a narrow road between crops

You should begin work on roadways in a course set in a large field. Figure 10.6 shows how to organize such a course. The roadway should be approximately 20 feet wide and should not lie along a fence line.

Begin by standing your dog near the mouth of the pen, but at a position outside it. This is the same position he held during flock conditioning (see Figure 9.1). With your helper standing beside you, call the flock out of the pen. The dog must stay in position until the last sheep has passed. Once this occurs, you may release the dog to patrol the roadway.

Previously you released the dog to patrol a one-sided boundary while you led the sheep. The dog patrolled one side of the flock while the fence line controlled the other. When fence lines control one side of a flock, it does not matter how strongly the dog presses the flock against it. It only matters that the dog stays on his side of the boundary line. On roadways, however, your dog must adjust to the flock. This means he must work farther away from the sheep along an invisible line located at the boundary of the sheep's' flight zone. This will enable him to keep the flock on the road.

To accomplish this, give the dog an "out" command to make him retreat across the boundary before returning to his patrol. If the dog is not instinctively sensitive to the flock's comfort zone give him a second "out" command to make him move a greater distance past the roadway. This distance will vary with the sensitivity of the sheep. The dog will have to move at least 15 feet from "heavy" (placid) sheep and an even greater distance from light (nervous) ones (see Figure 10.7).

As I said, the dog must move away from the flock until he reaches the boundary of their comfort zone. Once this is reached, the sheep will relax and walk along the roadway. If the dog violates the flock's comfort zone the sheep will refuse to move.

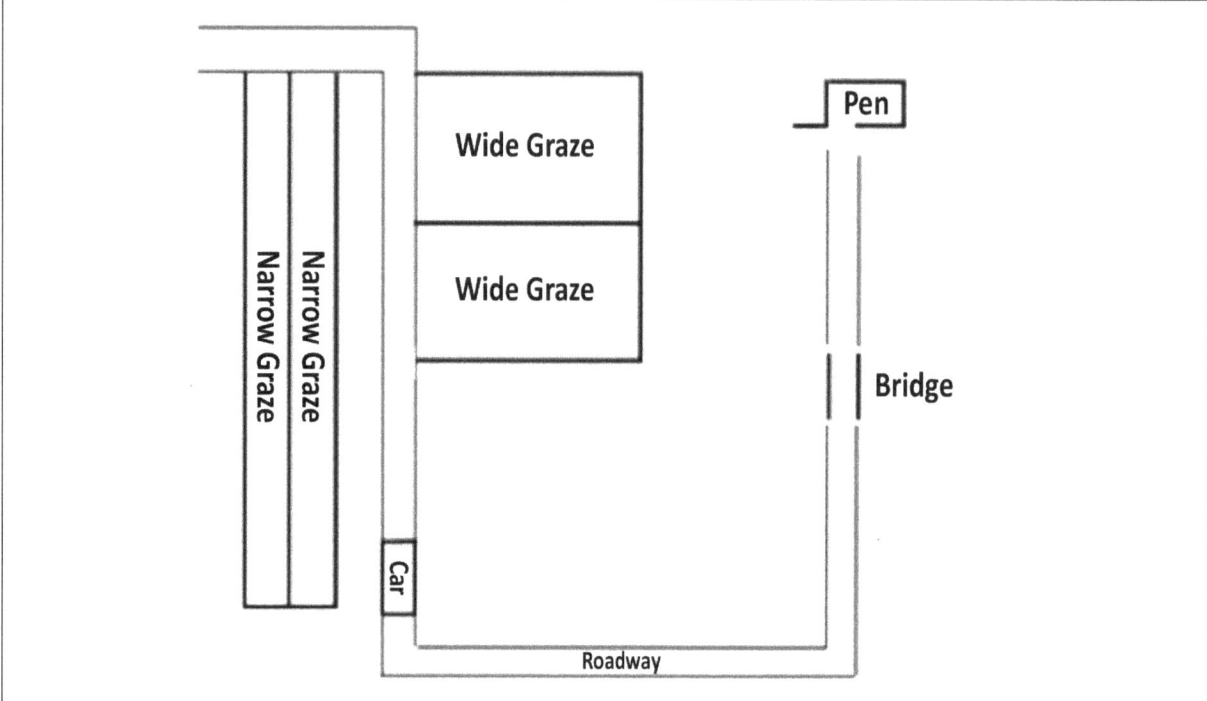

Figure 10.6

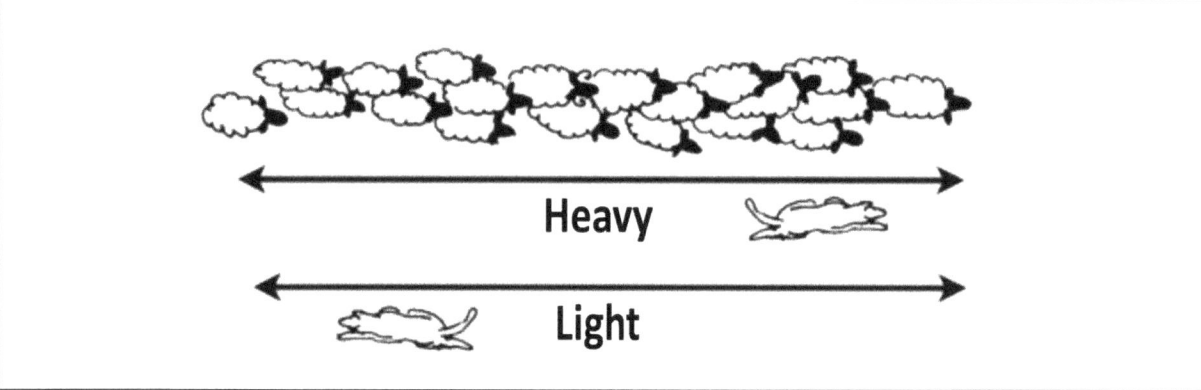

Figure 10.7

In contrast, the dog will loose contact and control of the flock if he patrols from too great a distance. As the dog gains experience working along roadways he will begin anticipating your "out" command and regulate himself along the proper invisible line. Your dog has begun to learn "balance" but finding the proper distance from the stock is only one element of "balance." Positioning himself at the proper point along the invisible line (which moves as the sheep move) is an equally important element. Only by finding the proper invisible line and moving along it to maintain the proper balance point, can your dog perform his three main tasks: to keep the flock on the road, to keep them from eating the forage alongside it, and to control the length and width of their column.

You might wonder how you teach your dog to find the correct balance point. The answer is simple. He learned to find that point when he learned how to traverse the flock in previous exercises (see Chapter 9). Balancing on the road is a little different but it will come with time. The talented dog will become frustrated when he is out of balance and will work harder to find the right balance point so he can perform his tasks properly.

If the flock is in the correct formation because the dog has found the balance point he does not always need to traverse the full length of the flock. He may take, for example, three shortened trips for every complete one.

You will notice that when the dog is balanced the sheep will not stray off the roadway. As we shall see later in the chapter, if traffic is present the flock must not occupy the entire roadway or trucks and automobiles cannot pass. Maintaining the proper balance point is difficult under these conditions, for not only must your dog escort vehicles past the flock, but he must also keep the flock in the proper position on the road. At such times you might wish to use two dogs, one to control the flock and the other to escort vehicles past it.

The dog will need to change sides when the flock encounters hazards. Use the "Switch" command for this. Make sure that the dog is far enough in front of you when asked to change sides. Use your previous knowledge of changing sides to do this.

Pressing the Flock

Once both you and your dog are proficient on roadways, begin working him along the sides of the flock in places without discernible boundaries. Although the majority of the dog's work will involve boundaries, sometimes he must work along the flock when there are no boundaries, e.g., in a pen, a wide graze or when traffic is in the road.

To begin this exercise return to the wide graze, but now work it like a narrow graze. Enter the wide graze and walk in front of the flock. The flock should keep in a narrow column as you lead them forward. Leave the dog outside the graze so that he can be called into a position along side and just to the left of the flock (see Figure 10.8). Then call your dog into the grazing area with a "walk on" (you learned this command as part of the "placement before the flock" exercise). This should be immediately followed by the command "come here." The dog should come directly to you alongside of the flock and not along the boundary line. Once he reaches you send him back along the flock with the "go on" command (see Figure 10.9). As the dog becomes comfortable moving along the flock, he will naturally start to move closer to it. As he does so, name the behavior "over right," but keep your voice as inviting as possible to encourage the dog to continue moving closer to the sheep. Allow the dog to press the flock to the right side of the grazing area from its present position. Although your dog may enjoy this exercise, don't allow him to enjoy it too much. That is to say, don't

Photograph 10.5 Pressing a flock of 1,000 head along a furrow on his left and using two dogs on his right.

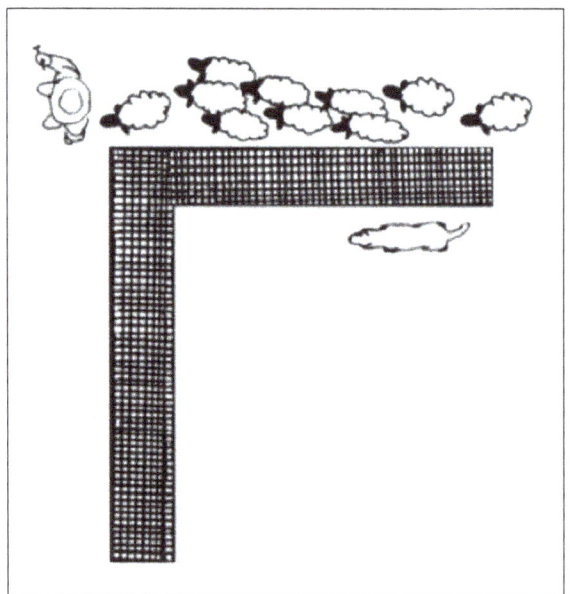

Figure 10.8

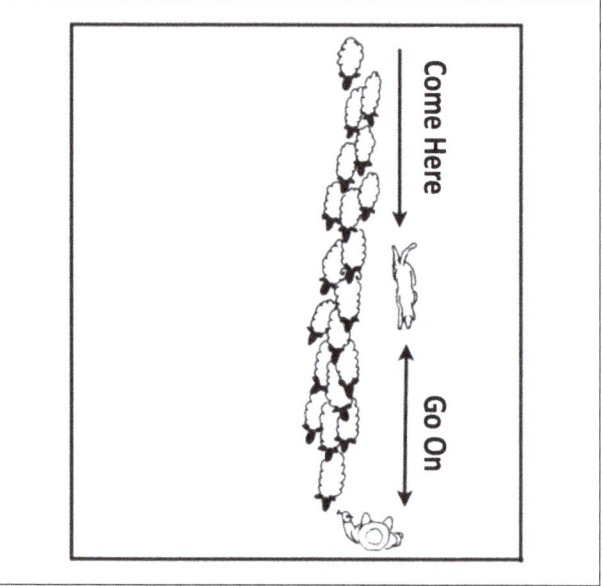

Figure 10.9

Chapter Ten: Advanced On-Stock Training

tolerate his gripping the sheep as he presses them to the right.

Earlier we discouraged the dog from working close to the flock. Now, by the command "over right" we allow him to work close to press the flock. You can also use this command to tighten the flock when you want to leave a wide graze area. In essence you are making the graze narrower by bringing the dog into the area.

Traffic

If the shepherd wishes to stop the flock on the road, or if he wishes to turn it into a side road, he does the same when they meet traffic on a narrow path; the dog must then elongate the flock by pressing it to one side and allowing only a few to pass at a time, thus making a way for the traffic. In such cases, the dog

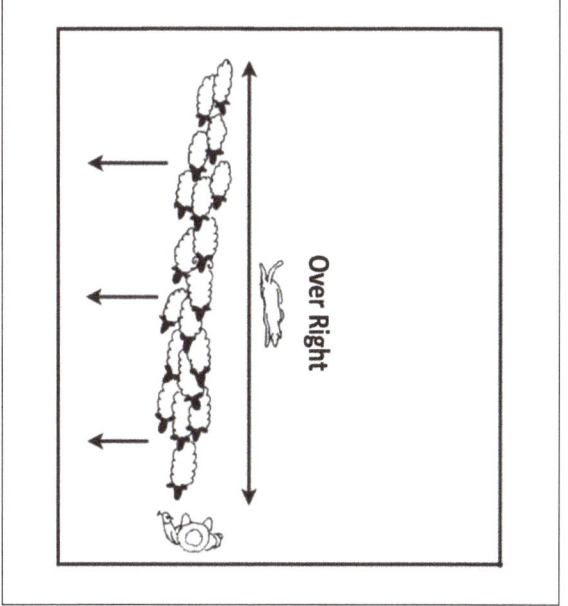

Figure 10.10

Photograph 10.6 During a competition this dog has already pressed the flock to the right side of the road and is now going to meet the car and then escort it past the flock.

must always keep between the sheep and the traffic, to prevent the sheep from running under the wheels. The dog therefore, must show no fear with regard to any vehicle, (not even a rattling motor lorry), nor of the driver when he cracks his whip (Von Stephanitz 1925:300).

The dog has now learned to move the flock along a roadway without pushing the sheep off the road and has mastered pressing the flock within a wide graze area. These form the foundation for escorting traffic past the flock. If you first move the flock to the right side of the road and then escort the vehicle past the flock, the flow of traffic will not be affected.

As the flock travels along the road filling its entire width, send the dog along the left side of the road. When traffic approaches, give the "over right" command to press the flock to the right hand side of the road and allow the traffic to flow past. In time the dog will associate this command with traffic work and uttering it will alert him to on-coming traffic (see Figure 10.11).

Begin the exercise with the car's engine off. Then increase the difficulty by starting the engine in neutral while leaving the car parked. As the dog gains experience set the vehicle in motion at a slow and steady rate. At the beginning, make sure the car gives the flock a wide birth so the dog has time to adjust to its movement without interfering with his work. As the dog becomes more proficient, move the auto closer to the flock. Cars should travel in both directions past the flock (see Figure 10.12).

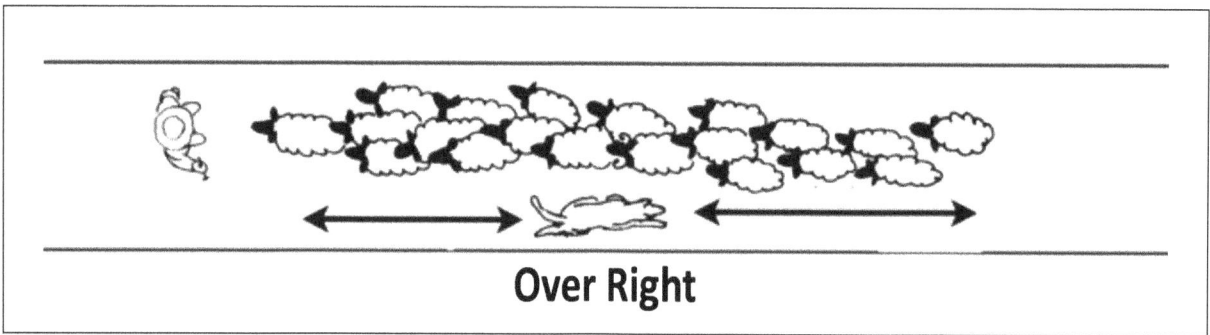

Figure 10.11

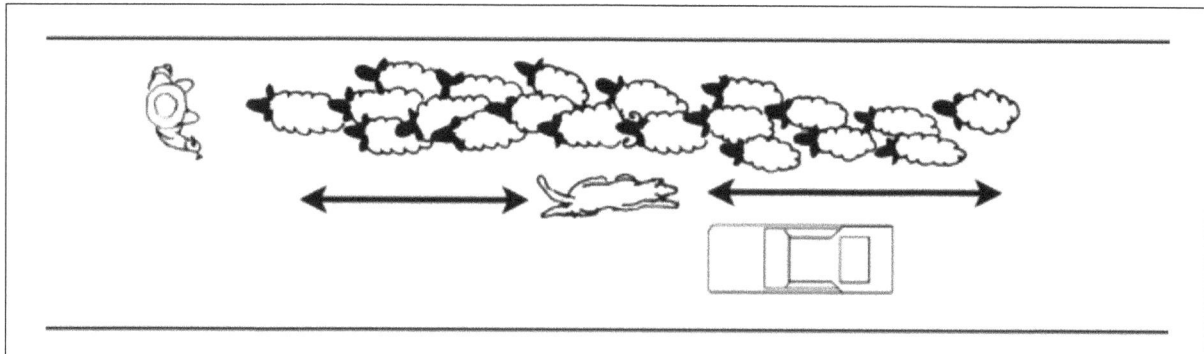

Figure 10.12

Chapter Ten: Advanced On-Stock Training

Encourage the dog to escort the car past the flock. To do this he should position himself at the car's front wheel, near its bumper, and escort the car past the flock from there. If a member of the flock is out of formation the dog will be in a position to correct it. You will find it easier to place the dog at its bumper when the car approaches from the front of the flock. From your place at the flock's head send the dog with the car as it approaches.

To teach the dog this part of his job, walk with him, showing him his correct position at the front wheel of the car. Use the "pick" command to accomplish this. When the dog first learned the "pick" command he was taught to look for a stationary object to go to. Now he must learn to go to a moving car. In time, the dog will hear the command and go directly there. And once positioned he will learn to move with the car. If needed repeat the "pick" command while the car is moving (see Figure 10.13).

Figure 10.13

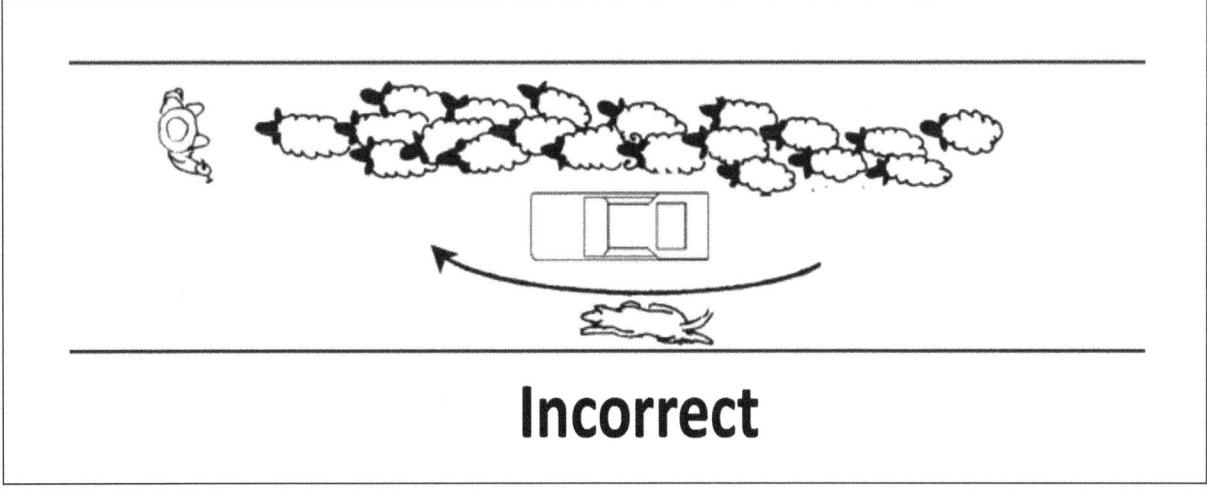

Figure 10.14

Remember, the dog should always be between the flock and the auto. He should not go around the vehicle and leave his charges unprotected from the oncoming traffic.

Flanking and fetching dogs (e.g., Border Collies) will have difficulty executing this exercise. First, if they control the flock by driving them from the rear, they will send the sheep across the edges of the road, but then, if they control the flock by flanking along its side, they will naturally include the car with the flock, thus leaving the flock unprotected from the car.

Bridge

Begin the bridge exercise by placing your dog on a "stand-stay" at a wing of the bridge. Use the "pick" command to position him correctly. The dog's chest should not extend past the plane of the wing panel.

In this position the dog can watch the front and the rear of the flock while it is traveling over the bridge. Place the dog on that side of the bridge where either the possibility of danger exists or where there are crops you do not want the flock to eat. From here the dog can passively control the situation.

Now put your dog on a long line. Go to the other end of the bridge and call him to you. The dog must not run around the bridge but pass over as if it were the only way to cross.

Next, place the dog on a "stand-stay" 100 feet away from the bridge. Then call the dog to the bridge, positioning him at the panel. Now repeat the previous step; put the dog on lead, walk to the end of the bridge and call the dog to you. Once the dog

Figure 10.15

performs the exercise consistently, proof him with the lead removed. When he has mastered the exercise without the stock, add the sheep.

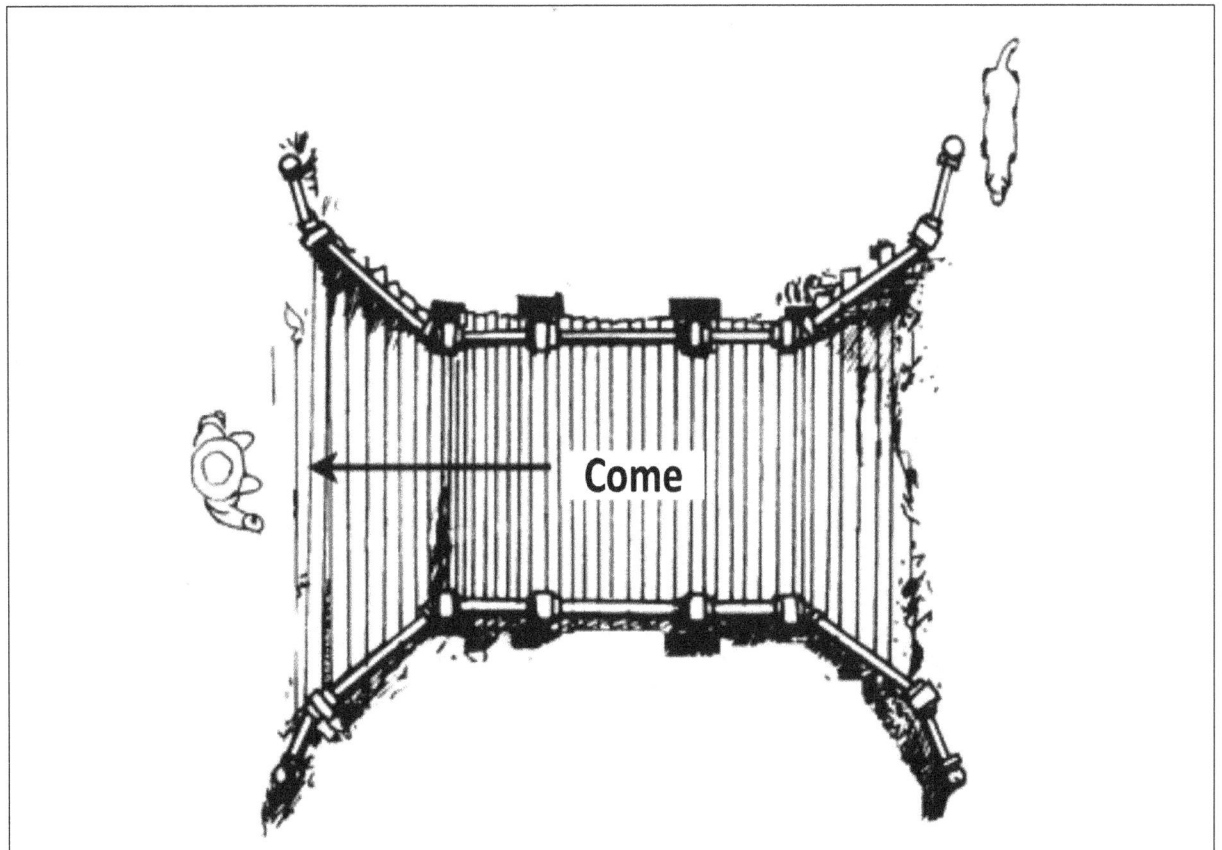

Figure 10.16

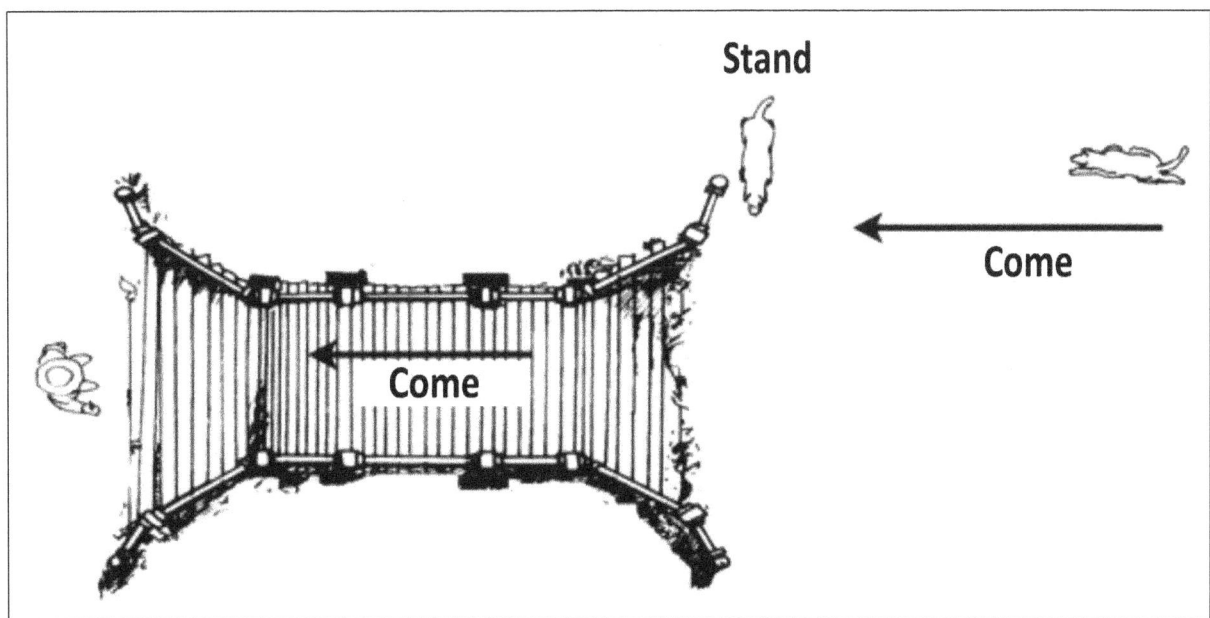

Figure 10.17

Bridge with Sheep

As you lead the flock toward a bridge, you will notice that one side is either more treacherous or more inviting to the sheep than the other. You must place your dog on that side to protect the flock from harm and/or temptation. Call the dog to you. As you approach the bridge position him at the wing on the dangerous (tempting) side using the "pick" command.

Your dog may be working on the left side of the flock, for example, while the dangerous side of the bridge is on the flock's right. In such cases you must position the dog on the opposite side (not end) of both the bridge and the flock. To accomplish this, call the dog to you and give him the "switch" command, thus moving him to the opposite side of both the flock and the bridge. Then ask him to stand guard. It is now his responsibility to protect the flock from danger while you lead the sheep over the bridge (see Figure 10.18). The dog should hold his position until the entire flock has safely crossed the bridge. He may, however, leave his position upon command if a situation develops which demands his presence elsewhere.

In many cases there will either be no water flowing under a bridge or in trial situations an obstacle representing a bridge with no stream underneath. At such times the flock, as well as the dog, may try to avoid crossing the bridge (or obstacle representing the bridge) by going around it. Insure against this by approaching the bridge at an angle. By approaching at an angle the flock leaders will be pressed against the side of the bridge where the dog is standing. Now the leaders will follow you across the bridge, and the rest of the flock will follow.

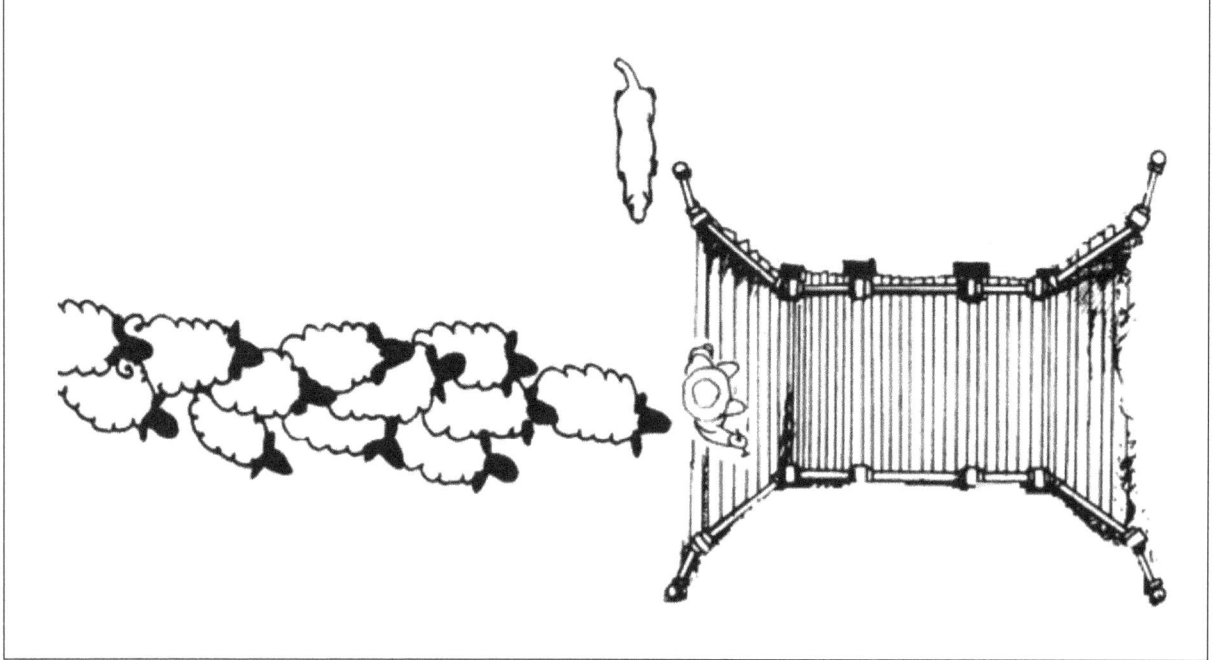

Figure 10.18

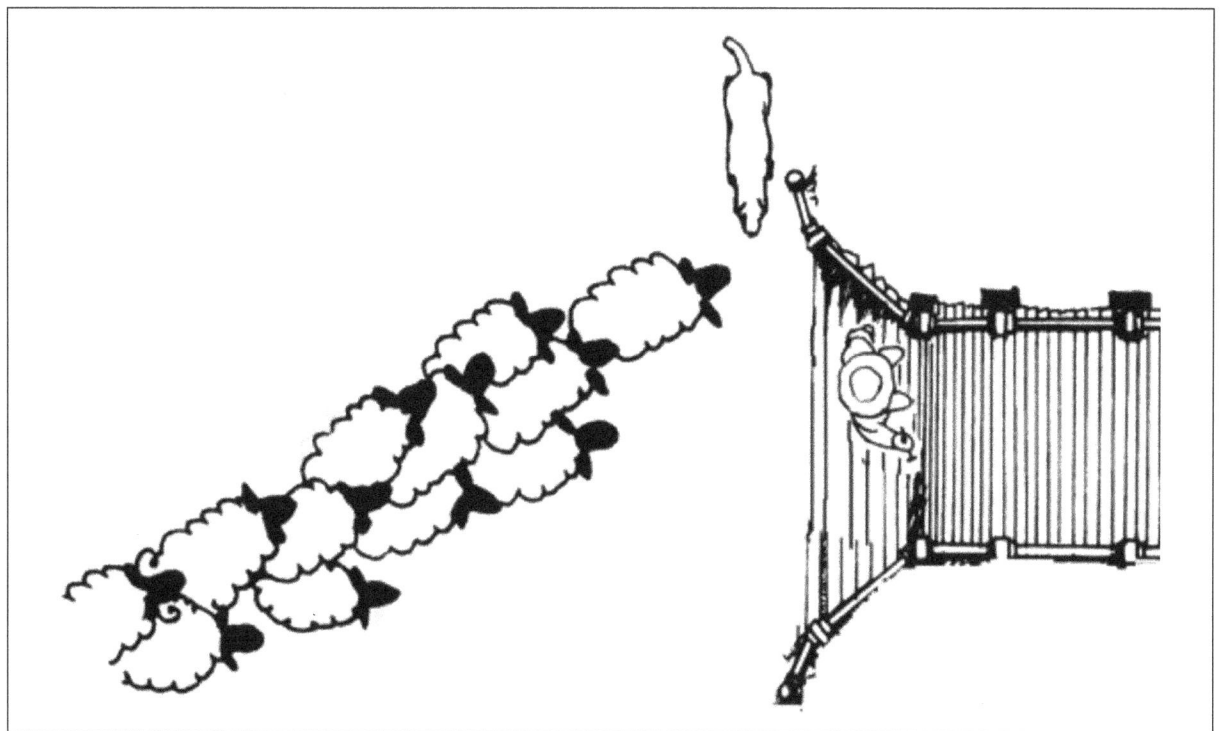

Figure 10.19

The Pen

There are many aspects to pen work, two of which include your relationship to the flock and jumping your dog into the pen. Both are equally important. In earlier exercises you laid the foundation for this work, i.e., in conditioning the stock to your dog, in working the two sided boundary and in teaching your dog to jump. Now we build on this foundation to teach your dog to move the sheep from the pen in a quiet, controlled manner.

The first step in the process is to develop a relationship with the flock (see sheep conditioning page 150). The next step is to teach your dog to jump into an empty pen. Finally, you teach the dog to jump into the pen with the sheep. If you have not taught your dog to jump already, refer to "Jumping" on page 138.

Start with a pen roughly 32-48 foot square. It should have an 8-12 foot gated opening through which the sheep can enter and exit. The pen should be constructed of material that allows the dog to see into it.

When you first approach the pen, position your dog with a "stand stay" in the mouth of the gate. You can keep your dog in position by driving your shepherd's staff into the ground and sliding your dog's lead over the end. Make sure he is close to the panels of the pen a foot or two from the inside. If he is not, then when the gate opens, the distance between the fencing and the dog will be too great. Then the sheep may escape before you want them to leave the pen. For this reason I prefer hinge-free sliding

Photograph 10.7 Two dogs guarding a pen opening.

gates. You can secure sliding gate panels in many ways. For example, in Germany one end of the panel is laid against a corner post while the other end is sandwiched between two other posts. Each competitor pounds these other posts into the ground. This anchors the panel at both ends.

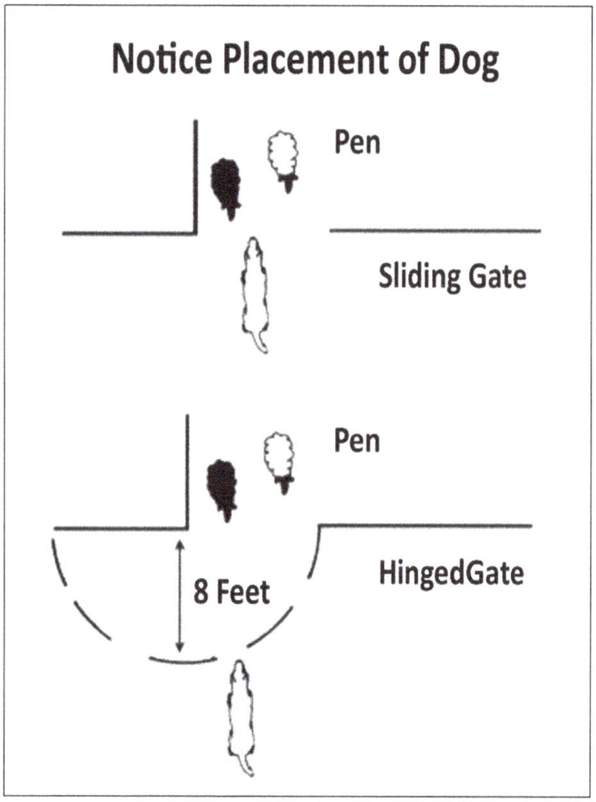

Now, you can slide back the gate panel and walk in. It is imperative you gain the trust of the flock as soon as possible. To do this talk quietly, walk slowly and feed the flock, especially the lead sheep. Once you have the flock's trust, return to your dog and send him into the pen.

The dog should jump into the pen at the location shown in Figure 10.21. It is important that the dog jump at the point where he will press the flock along the wall of the pen opposite the side noted as A in Figure 10.21. If the dog jumps into the pen at the side in which the opening is located, he will press the flock to the back of the pen.

Figure 10.21

In Germany competition flocks number between 100-300, but are placed in pens of approximately twice the size as those used in AKC "C" course competitions. In these

Figure 10.20

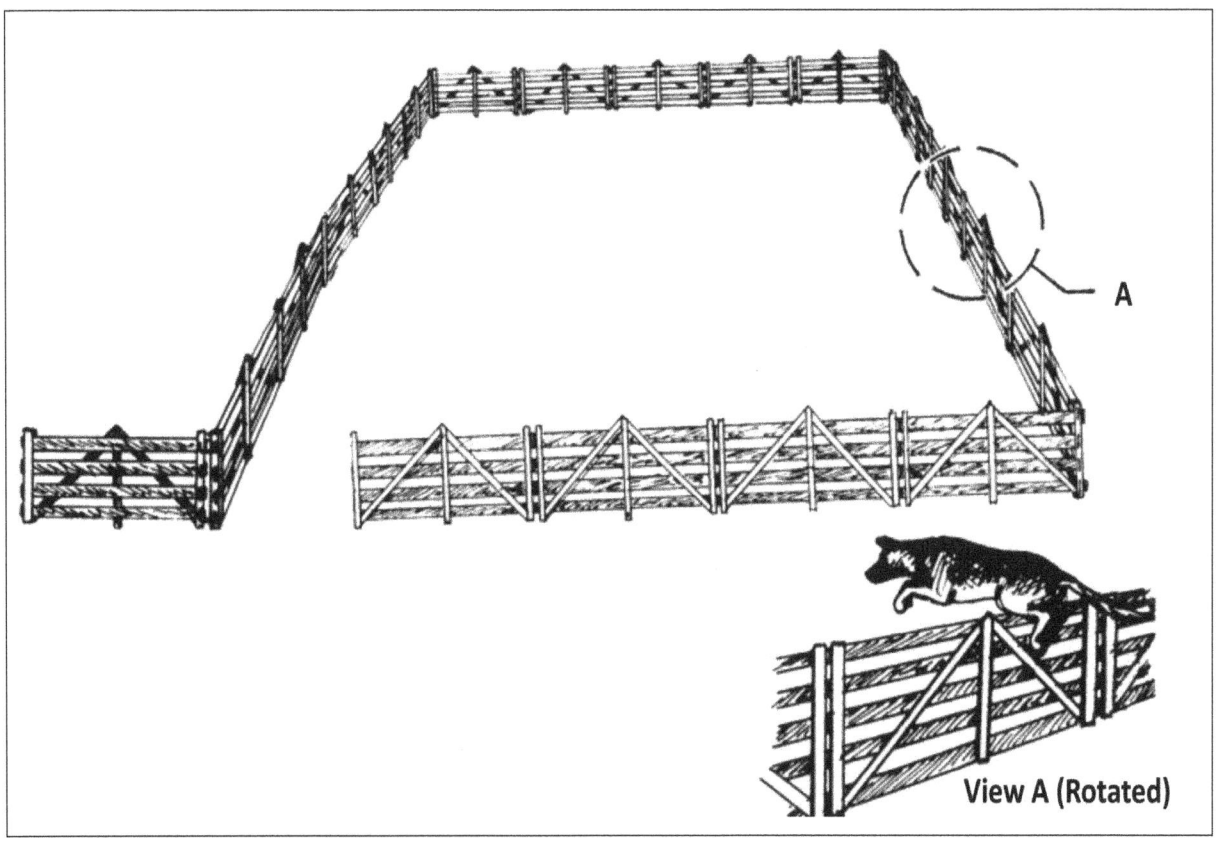

Figure 10-21

cases, the dog must jump in wherever he can find room, which is generally at the corner.

At the beginning of training, the panel in the fence where the dog jumps in should be only 1-2 feet in height. As the dog gains proficiency, gradually raise the panel until it is approximately 33 inches high, the height of the rest of the fencing. At this height the dog may use the top of the panel to push off as he jumps, although clearing it cleanly is preferred. Be sure to cover the panel with a sheet in early training. If the dog can see open areas in the panel, he might try to jump through them, especially if the obstacle is high.

As the dog jumps over the panel into the pen give him the command "stand stay" to settle him. In the future when sheep are in the pen the dog must land squarely on his feet with a minimum of movement to decrease stress on the flock. The "stand stay" command will help him learn this important lesson.

As the dog becomes proficient in jumping into the pen you should gradually make your way to the gate (see Figure 10.23) and call the dog over the panel from there. Once you can call the dog over the fence, command him to the outside edge of the pen's mouth using the "pick" command (see Figure 10.24). The dog must be in this position to protect the sheep from injuring themselves as they exit.

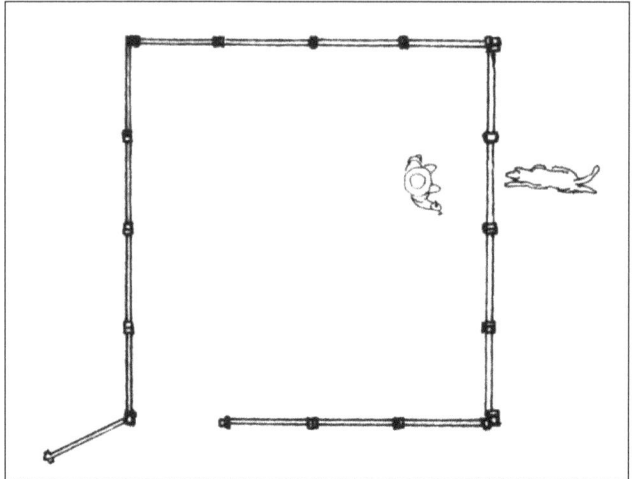

Figure 10-22

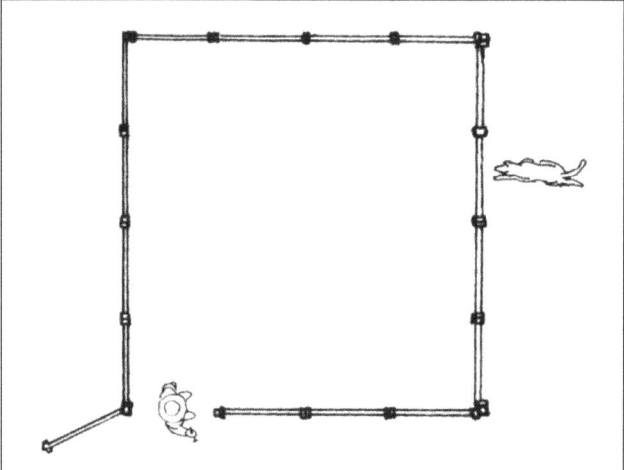

Figure 10-23

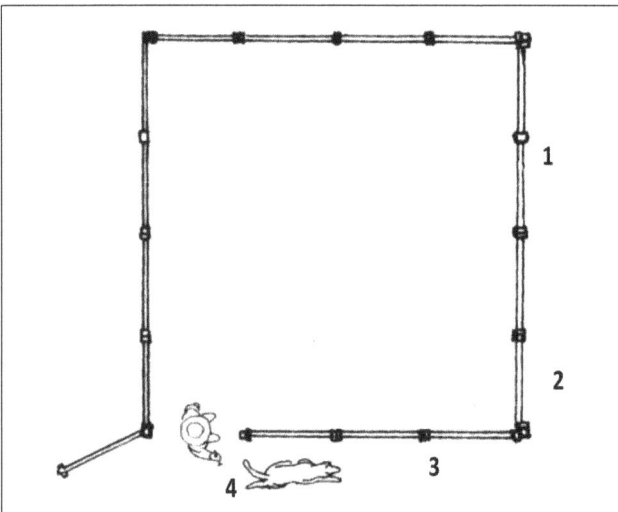

Figure 10-24

Earlier you placed the dog in a stationary position at the place where he jumped into the pen. Now it is time to start him at your side. First, place your dog at different locations along the pen. This will teach him to run farther away from you each time before jumping the fence. Now send the dog with the following sequence of commands: a "go on," then a "jump," then a "stand stay," and finally, a "pick" at the opening of the pen.

At a trial this chain forms one fluid motion. Yet when teaching it, break it into its elements and teach each part separately but think about teaching them backwards. Work on the stand at the corner first. Then, add the stand after jumping into the pen. Next, add the jump, etc.

By doing the series backwards, the dog is reinforced by the part of the exercise he already knows. This will shorten the learning time. Then, when your dog has learned all the parts, put them together. When the training for this exercise is completed, you can send your dog anywhere on the pen's perimeter, jump him into the pen, stand him motionless and then move him to the opening of the pen to oversee the flock's exit.

We are now ready for the final exercise you and your dog must master. It is time to jump the dog into the pen with the sheep. Because of the flock's position in the pen the dog may become distracted as he attempts to jump in. He may not have the same level of confidence he did without the sheep present. Consequently, you should lower the height of the fencing. Now repeat the previous exercises with the livestock in the pen.

In both German and American trials the dog may enter the pen through its mouth without jumping in, although entering the pen via jumping is preferred. The advantage to jumping is the element of surprise the dog achieves as he leaps into the pen. This helps disperse the sheep evenly to a place from where they can easily exit the pen. In contrast, entering through the mouth presses sheep toward the back of the pen. This gives the sheep mixed signals and they are no longer sure they should leave the pen through its mouth.

It should be obvious that where the dog enters the pen is critically important. Observe your flock before you send your dog to jump into the pen. Look for the point of balance that will cause the flock to disperse evenly to the proper locations. In most cases this will be at the middle of the flock.

As mentioned earlier, you can choose to send your dog into the pen through the gate. Nevertheless, he should go to the same place he would have found himself had he jumped in. When he reaches that point, call to the flock. As soon as the flock heeds the call and begins to follow you, call your dog to his "pick" position at the mouth of the pen. Timing here is important: Your dog's movement and positioning control the length and width of the flock as they leave the pen. The dog should cause a funneling effect as the flock leaves the pen (see Figure 10.29). When you return the sheep to the pen at the end of the session, post your dog at one side of the gate while you go to the other. This will create the same

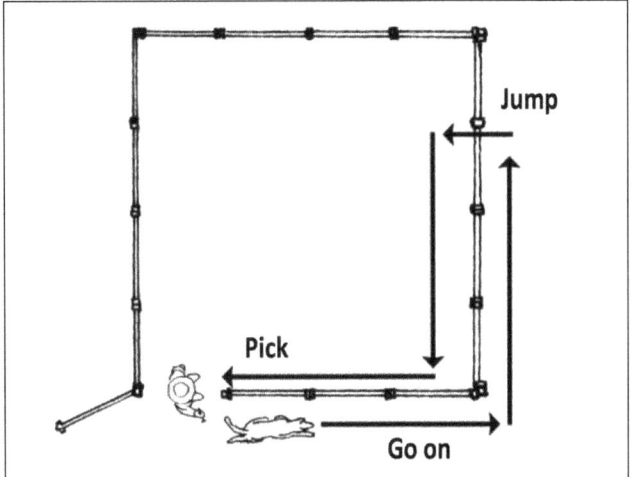

Figure 10-25

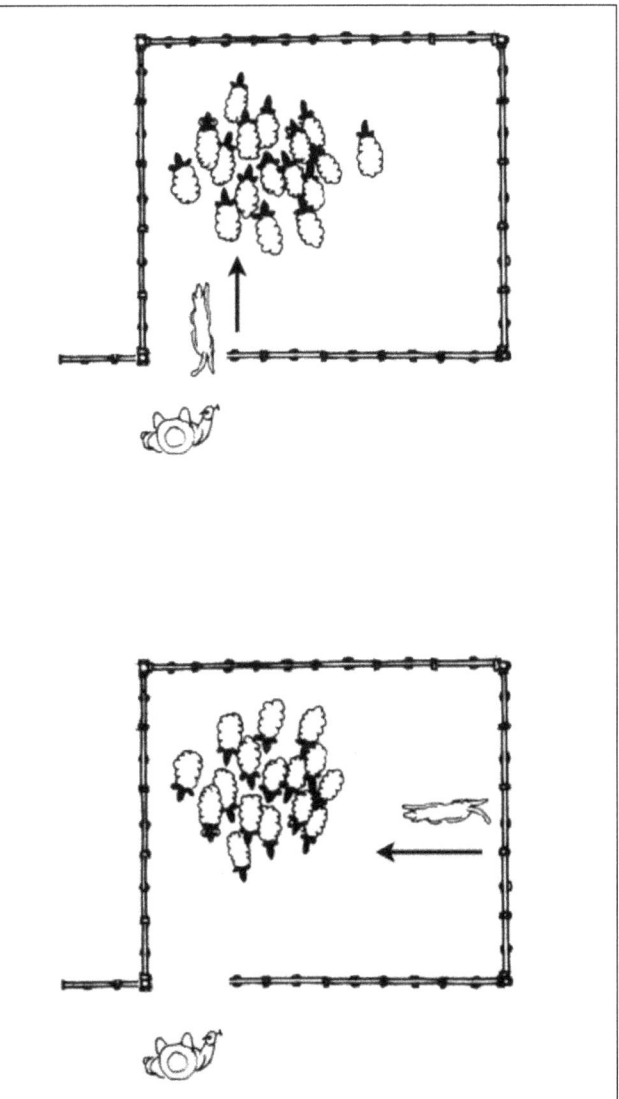

Figure 10-26

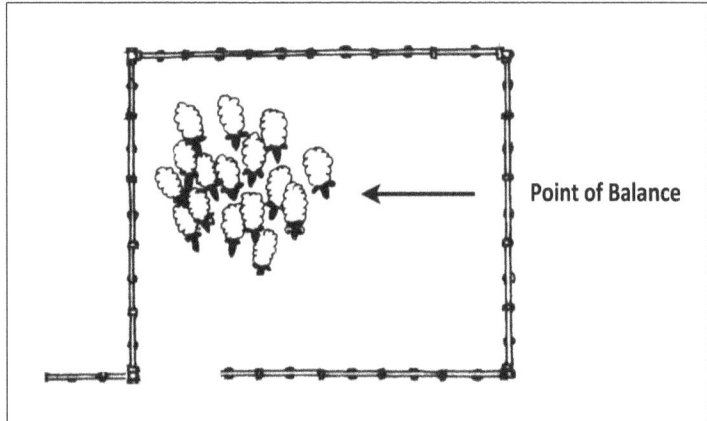

funneling effect you created when the flock exited the pen. As a result, you will find it easier to count the sheep and look for injuries (see Figure 10.30).

Figure 10-27

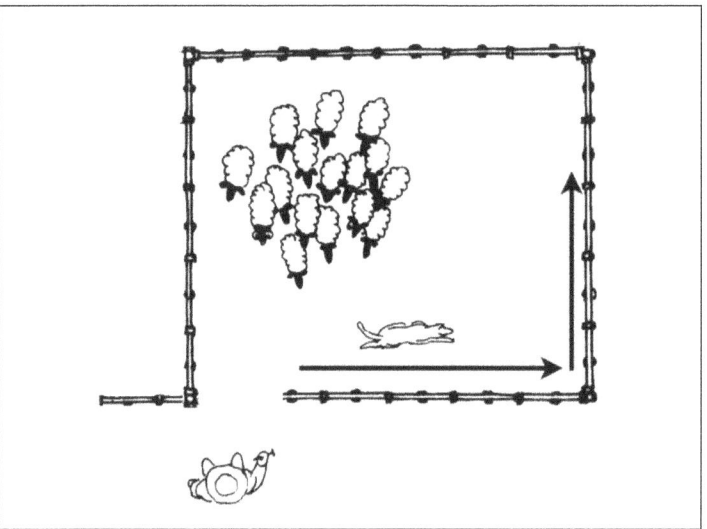

Figure 10-28

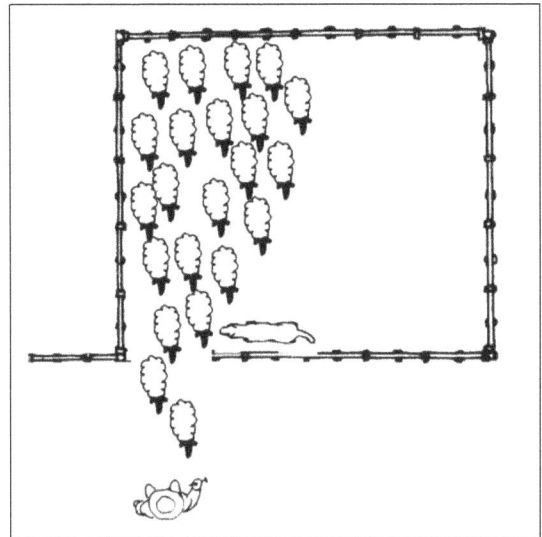

Figure 10-29

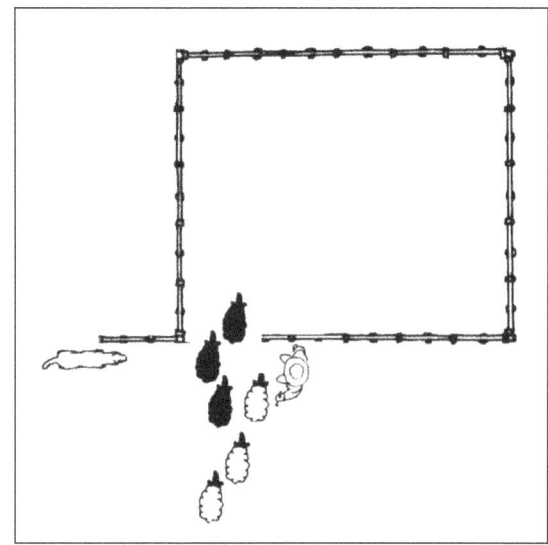

Figure 10-30

Competitions

Originally, herding competitions tested whose dog could best perform daily shepherding routines, thus providing breeders with information useful in improving the structural and temperamental quality of their dogs (Von Stephanitz 1925:316-317 and Scientific American: May 7, 1898). Americans are most familiar with those competitions that require controlled movement of stock from place to place, e.g., International Sheep Dog Society Trials and the arena trials of the AKC, AHBA and ASCA. The classic *Bob Son of Battle* (1898), which many of us grew up reading, featured such trials. It described a rivalry between two neighboring British shepherds battling over the age old question, "whose dog is best?"

During the same year that *Bob Son of Battle* first appeared in print, the Scientific American (May 7, 1898) published an account of a very different kind of herding trial held in France. Three years later, Germany held the first SV tending competitions (Von Stephanitz 1925:316-317). These trials included well thought out plans for nourishing stock as part of the competition. For tending shepherds were primarily concerned with optimizing their sheep's food intake, thus insuring healthy animals.

Competitors in herding trials in Great Britain, France and Germany often have to be full time shepherds. In the past, some participants even needed letters from employers to enter competitions. In contrast, today's competitors come from different occupational, professional and lifestyle backgrounds. Often their aim is to simply enjoy a weekend of competition. Consequently, many competitors do not own their own sheep and some go to training centers for only an hour or so each week, leaving the feeding and care of the animals to their owner.

To successfully compete in tending competitions you must know two things: first, the best sheep nutrition and second, how to move a flock from place to place. Thus, tending competitions are distinctive because they require handlers to have an uncommon expertise in animal and plant science, as well as dog training and handling ability.

Only in a harmonious effort between man and animals, can optimal success in grazing be reached. The sheep may not be unnecessarily disturbed during feeding; they must be contained in certain boarders and are to be guided properly on streets and roads (*Hüteordnung für SV - Leistungshüten* January 1989).

The overall feeling of tending competitions should be pastoral bliss. The handler should conduct himself and his dog in a quiet, controlled and professional manner, using audio and visual signs to control both dog and flock. The dog should respond to the handler with quickness and a willingness to please, while the dog's work in relation to the stock should be natural and without regular direction from the handler. The flock should show respect and trust for both the handler and the dog.

In European competitions shepherds sometimes have two dogs working the course. Specifically, in German trials (SV and AAH) many competitors use two dogs, but only one is judged. The second dog's job is to make the run smoother by performing the tasks of a second dog in everyday herding. This relieves the handler from performing these same tasks. In contrast, French Trials (SCC) allow handlers to use only one dog.

The number of sheep varies from competition to competition, ranging from 50 - 300. In the United States, many clubs are hard-pressed to find flocks larger than 100. Although clubs may need to experiment before determining the optimum number, I imagine that somewhere between 50-100 sheep is most realistic for us.

Three tending organizations[34] hold trials in France and Germany. The trials held in Germany are sponsored by the SV (Verein Für Deutsche Schäferhunde) and AAH (Arbeitsgemeinschaft zur Zucht Altdeutscher Hütehunde). Both German organizations hold trials under the same rules. The SV trial is only open to German Shepherd Dogs while the AAH is open to all breeds. The trials in France are held under the auspices of the Société Centrale Canine (SCC) and are open to all breeds.

In the North America trailers have several options as well: The American Kennel Club (AKC) offers tending trials on what is called the "C" Course; the American Herding Breeds Association (AHBA) sponsors tending trails under French rules; and Shutzhund USA (under the auspices of the SV) and the American Tending Breeds Association (ATBA) sponsors competitions under SV regulations. In Canada competitions are held under the Canadian Kennel Club (CKC) regulations. The Australian National Kennel Council (ANKC) and the Federation Cynologique Internationale (FCI) also offer tending competitions.

Differences exist in the levels of competition and requirements for certificates and titles. For instance, the German course has only one level of competition while the French, ANKC and AKC courses have three. A dog must only qualify once to earn its

---

34 A fourth organization (Continental International) has held tending type trials in the past (the first was held in 1985). The trial moves from country to country each year and so may take a different form (tending/I.S.D.S.) depending on where it is held.

German Working Certificate (HGH). In contrast, dogs participating in French trials earn certificates after obtaining one passing score at second and third level trials and a championship title (CACT) after obtaining a total of three passing scores at level three competitions. The AKC awards three titles (HS, HI and HX) and a championship title (H.Ch.).

Sheep are generally used on all the above courses but the SCC also holds cattle trails on the same course and under the same rules. When competing on a SCC course, the flocks (50-80 head) may be used 2-4 times. The SV runs a flock (100-300 head) approximately six times, using two groups per day of competition. The AKC "C" course allows sponsoring clubs to use groups of sheep (minimum of 20 head) 6-8 times. The more times a flock is used the more accustomed to the course its members get. This can be both advantageous and disadvantageous depending on the stock and the dog.

You might think it difficult to prepare for so many different competitions, but is not so. Tending trials have many exercises in common. Each begins and ends at a holding pen and in each the flock is led along roadways, past traffic and to a grazing area, as Alain Pecoult wrote in the *Working Sheepdog News*.

To begin with the handler sends his dog inside the pen or barn to bring the sheep out as he calls them and starts walking away, the dog driving the flock behind him. Courses vary, but every course will have the following obstacles in whatever order they may appear. From the starting pen the sheep are driven to a pasture. On the way they have to go through a narrow gate or over a bridge in an orderly way and pass along some appetizing crop or pasture. Of course, the dog must not allow them to stray off the road or dirt track they're walking on. When they reach the pasture, the shepherd and his dog take the sheep to a part of it which is marked by four posts erected at corners and about twenty yards apart (there are usually two or three such squares). The sheep must be left to graze undisturbed inside the square for two or three minutes, and of course they mustn't come out of the square. Then they're off again and back onto the road or dirt track. Before getting there the dog must stop while the shepherd makes sure it is safe to walk on. Then the sheep are made to move on again and, as they're walking on the road, a car must be allowed to pass the flock without stopping. It may come from the rear or front. Eventually the sheep are brought back to their starting point; they must be penned and sometimes the shepherd is asked to leave his dog to guard the entrance to the pen while he walks away (1979).

Differences in competitions lie in the rules governing the way in which the dog/handler teams can accomplish these tasks. For example, in French trials the handler and dog may take any position during the course of the competition and the dog may drive or fetch the flock through the course. In contrast, handlers competing under German rules must walk in front of the flock while leading them along a roadway. While they are in the narrow graze they must walk with the last third of the flock and dogs must change sides in front of the flock rather than behind it. Because I have concentrated on the German style of tending in this book, I will begin by discussing German tending competitions. The chapter ends with a discussion of French, American Kennel Club and International FCI trials.

German Trials (SV)

Exiting Out of the Pen (SV 6 Points)

During competitions much happens before the start of the clock, which sets the mood for the entire run. The first contact with the sheep takes place at the approach to the pen. Thus, handlers should take time to walk around the outside of the pen with their dogs.

Once this is done, the handler opens the gate and stations the dog in front of it. The handler's staff is an excellent tool to use in correctly positioning the dog. You should place the shovel in the dirt and then slip the leash over the stick, thus tethering the dog in place.

Position the dog close enough to the gate to keep the sheep from escaping. Then, enter the pen to further familiarize yourself with the flock. The bond of trust established here can make or break your performance. As you walk among the flock seek out the dominant sheep; she will be the lure that the flock will follow around the course. Offer her something to eat to develop a trusting relationship. As she begins to

Photograph 11.1 Dogs tethered in place

Photograph 11.2 As the flock exits the dog stays in place until the last sheep leaves the pen.

trust you, slowly make your way to the open gate as she follows. Leave her in position at the gate to ensure a timely exit from the pen.

Until now you can't loose points. This changes, however, as the dog leaves your vicinity. Then the judging begins. Handlers have two options regarding the way their dogs enter the pen; they can jump in or enter through the open gate. In either case, dogs should wait until their handlers give the command to enter. If a dog chooses to enter without command, the team will lose 1/2 to a full point.

Once a dog enters the pen he should take a position enabling him to press the flock against the wall of the pen. Now, handlers should call or coax the flock to leave. If the flock does not begin to follow a handler, his/her dog must slowly apply pressure to the flock, encouraging the sheep to exit smoothly. If the exit is not accomplished in a smooth manner, the team will lose up to 2 points.

As the flock leaves the pen, handlers should call their dogs to a post position at the opening of the pen. This changes the shape of the exiting flock; it narrows at the gate. The result is a funneling effect which ensures an orderly exit from the pen. Handlers can loose 2-3 points if this is not accomplished in an orderly manner.

Dogs should stay in a sentry position while watching over the flock as it exits. If they leave their position too soon, they will disrupt the flow of the flock. This can cost handlers up to 2 points. The ideal position here is for the dog's chest to be even with the opening of the gate. Being out of position can cost up to 2 more points.

If, however, the flock (or part of the flock) moves in an undesired direction during the exit process, handlers may call their dogs from their post to aid in changing this direction with no point deductions. If the dog leaves the pen too soon, the team will lose at least 2 points.

Hear My Voice: An Old World Approach to Herding

Photograph 11.3 300 head slowly moving along a narrow road.

NARROW ROAD (SV 8 Points)

Once you lead your flock from the fold, you begin your journey to a grazing area via a roadway. You should keep the flock on the road, not allowing your sheep to spread out along the road's grassy edges. You should also pace the flock for the journey. If you allow the flock to run ahead of you, they may tire long before they arrive at the grazing area. The pace of the flock should be about 2 miles an hour. You should adjust this speed according to the members of the flock. Ewes heavy in lamb or with lambs at their sides will need to move at a slower pace. You should also consider the climate or temperature of the day. All of these variables can affect how well your flock performs.

Some judges will deduct up to 3 points when the flock does not move in an orderly manner. This ultimately depends on the expertise of the handler. Shepherds control the speed and direction of their charges while their dogs control the length and width of the column. While traveling along a road dogs should work along the side of their flocks where the most danger or pressure to leave the road exists. Dogs should work without command, patrolling the entire length of the column. If a dog does not traverse the length of the flock, the team may lose up to 2 points. Failure to cover the entire flock may cause it to split into two smaller groups. If this occurs, the team will lose an additional 3 points.

When handlers see that their dog needs to work the opposite side of the flock, they should call it to the head of the flock, thus ensuring a smooth cross to the other side. If dogs are allowed to cross behind the flock they will pressure it, causing the sheep first to speed up and then to fan over the edges of the road. Some judges deduct as many as 4 points for this blunder.

When changing sides, the dog should not cross between the handler and the flock. If this occurs the team may lose 1-2 points. When changing sides, dogs should make right angle turns; if they arc the turn judges may deduct 1 point.

Handlers may need to stop their flock at some point while traveling to or from the graze. Perhaps they encounter a cross road or some other hazard. Because handlers control the direction and speed, they also control starting and stopping the flock. A well-conditioned flock will watch their shepherd. As he slows or accelerates, so does the flock.

As the handler slows to a stop, he may need the dog to help position the flock so that the sheep pass the hazard safely. Starting and stopping a flock is of paramount importance in accomplishing this. Therefore, not being able to do so is a major deduction, 2-3 points. Some judges deduct up to 2 additional points if a handler is too far away from his flock when calling them forward.

TRAFFIC (SV 10 Points)

As handlers lead their flocks along the road section of the course a vehicle will approach. Dogs should push the flock to the right hand side of the road to allow the vehicle to safely pass. To accomplish this, dogs should position themselves between the vehicle (near the front wheel) and the flock, thereby creating space between the

Photograph 11.4 **Dog escorting a car past the flock.**

Photograph 11.5 **Incorrect passage: Notice that the flock was not kept on the roadway because the second dog is at the front of the flock not working the other side.**

animals and the moving vehicle. Handlers should always position their dogs between the flock and any hazard to protect the flock from possible injury.

If the approaching auto must stop to avoid hitting the flock judges deduct 3 points. Similarly, judges deduct up to 2 points if the car must slow down to avoid the flock. Judges will deduct 1/2 to a full point if a dog does not maintain a position between the flock and the vehicle.

Judges deduct up to 2 points if the flock leaves the road while the automobile is passing by. Care must be given to protect the property alongside the roadway from the sheep. A second dog comes in very handy here. He can protect the roadside while the main dog escorts the car past the flock. Whether or not handlers use a second dog, the flock needs to stay on the road.

BRIDGE (SV 6 Points)

There are two types of bridges: natural and artificial. The first type of obstacle is the easiest to navigate. As you approach a natural bridge, the hazardous side is obvious to the handler, the dog and the sheep as well. The flock will be more willing to follow you over the bridge because there is no other way to proceed.

When approaching an artificial bridge handler skill becomes very important. Generally, this obstacle is in the middle of nowhere and nothing suggests to the sheep and dog that this obstacle is indeed a bridge (e.g., the "bridge" does not span a chasm or pass over a stream). Thus neither the dog nor the flock will naturally cross this obstacle, for they will see no reason for not going around it.

Handlers should position their dogs so their chests are even with the edge of the bridge (forming additional wings to the bridge). The dog acts as a sentry overseeing both the head and tail of the flock. After the flock crosses the bridge, the dog follows.

Photograph 11.6 Notice that the dog is chest level with the edge of the bridge.

Photograph 11.7 The dogs are first positioned at a distance and then moved closer to carefully balance the flock as it crosses over the bridge. Notice the angle from the pen to the bridge.

Judges subtract up to 2 points for each of the following faults: the dog does not stay in position, either moving about or leaving its post too early; the dog disturbs the flock while it enters or leaves the bridge; the dog is not even with the edge of the obstacle.

Judges will deduct up to 2 points if handlers don't position their dogs on the dangerous side of the bridge. Here again the use of a second dog is very helpful in properly directing the flock.

Handlers will lose all points for the bridge exercise unless some of the flock crosses over. If, however, part of the flock does not cross, a corresponding number of points will be lost. For example, if 1/4 of the flock does not cross the bridge, judges usually deduct 1/4 of the total points allowed for the exercise. If the dog does not cross, judges will deduct up to 2 points. If the dog has to be redirected to cross the bridge, judges will take off 1 point. If the dog does not wait for the last sheep to cross the bridge but leaves too early, the team loses up to 2 points.

In some instances, a flock crosses the bridge twice. In this case some judges score both passes of the flock on the bridge, but divide the total in half for each pass. Alternatively, they may score one pass as bridge work and the other as road work.

Photograph 11.8 Both dogs beginning to move as the last sheep cross the bridge.

WIDE GRAZE (SV 10 Points)

As the handler exits the road to graze his flock, the flock should follow him. Handlers can use their dogs as sentries while they walk into the grazing area. Once the flock knows it is time to graze, the handler should take a position across from his dog. This gives him an opportunity to observe the flock as they enter the graze. Generally, during the wide graze the handler does not have to accompany his flock as they graze. He must still, however, make sure that his flock receives the best nutrition possible. Because other flocks may have previously grazed the area during the competition, the handler and judge should carefully watch the condition of the forage.

The handler's knowledge of sheep and grazing is examined during this exercise. As judges observe a flock grazing, they look for one of three natural grazing patterns (see page 22). If the flock grazes according to one of these patterns they know the sheep are foraging as effectively as possible. When this occurs, handlers receive a full score.

If, however, judges don't see one of the three patterns, they look for causes. Is there too much pressure from the dog on the flock? If so, judges may deduct as many as half of the points available. If the dog hasn't caused the problem, then perhaps it lies with the grass itself. Has manure overly soiled it? Are the types of plants unpalatable to sheep?

If handlers are aware of these conditions during their runs, they should adjust their flocks accordingly. They should either maneuver their flocks within the graze or ask judges for a new grazing area. If they ignore such conditions, judges may deduct points.

The dogs' work is to protect the edges of the grazing area. They should hold these lines cleanly. If a dog is too wide from the boundary, most judges will deduct more points than if that dog is inside the boundary. Better to keep the flock from straying over the line onto forbidden grazing than to cut the grazing area by a foot or two. For example, a judge might deduct 2 points for being inside the line but 3 points for being outside it.

When running in the furrow, the dog must keep his hind legs in the furrow when turning. If he does not he can lose up to 2 points. The dog's task is to oversee the boundary line himself without direction from the handler. He should know exactly where the flock presses toward the boundary. Judges may deduct up to 2 points if handlers must tell their dogs where the problem lies.

Judges will deduct a significant number of points if the dog mechanically orbits the grazing area without regard for the stock. Generally, dogs take a position working the two boundary lines closest to the grazing sheep to contain them. Although an orbiting dog does contain his flock within the grazing area, he also burns energy for no reason. Dogs must be able to do a full day's work, but they cannot do so if they expend too much energy orbiting. Judges may deduct one third of the points (i.e. 3 points) for this fault. The team will lose fewer points, however, if the handler attempts to correct the situation. If the dog lies down, sniffs, or sits while on duty, a judge might hit the team with a 2-3 point deduction.

Photograph 11.9 **Dogs used as sentries as the flock leaves a wide grazing area.**

PLACEMENT BEFORE THE FLOCK (SV 8 Points)

This exercise should be carried out naturally during the course of grazing. Once the flock enters the grazing area, the dog should run ahead of the flock. By doing so, he minimizes the flock's damage to the forage. When entering a new grazing area most sheep will act like kids in a candy shop, sampling a little as they go, especially if the number of sheep is small. This wastes good forage. To prevent such damage, handlers command their dogs to enter the grazing area ahead of the flock, thus stopping its rapid forward motion.

To place their dog before the flock, handlers send them from the border of the grazing area to a distant position. The dog may be sent along the border of the pasture currently grazed or along a more distant border. The goal here is to give the dog adequate distance to approach the flock and slow its forward motion. If handlers are not able to send their dogs to this position, they will lose as many as 4 to 8 points.

Handlers must then ask their dogs to move in toward the head of the flock. If, however, they move in at the side or tail, judges generally deduct half the points. This is considered handler error. The dog should enter the grazing area at the direction of the handler. If the dog does not wait for a command to enter the grazing area, the team will loose up to 2 points. The dog must turn into the grazing area at a right angle. Failure to do so will cost the team up to 2 points.

As the dog approaches the flock, a handler must stop his forward motion at least three times. A team will lose one half to a full point for each time a dog fails to stop. During this time a dog's attention should be on his flock. If, on the contrary, it is on the handler the team can lose up to 2 points.

The object is to bring the dog as close to the flock as possible without disturbing its members. Dogs should move slowly and cautiously to smoothly accomplish this task. A lack of smoothness can lead to a one point loss.

The dog should also approach his flock with confidence. If he does not, judges may deduct one point. As the dog approaches the tip of the flock, the sheep should take notice and stop grazing. They should then calmly turn away from the dog and he must then hold his position. The team may loose two to three points if the flock does not perform in this manner.

Photograph 11.10 This dog has been sent to the heads of the sheep to begin the placement before the flock.

Once their flock turns away, handlers direct their dogs to leave the grazing area via the same route they took to enter it. Failing to return via the same route will cost at least 2 points. Once back outside the grazing area, the dog resumes his regular duties.

Photograph 11.11 Notice that this dog has made his stop just in front of the sheep without disturbing them.

Photograph 11.12 A dog exiting the placement along the same path it came in on.

NARROW GRAZE (SV 10 POINTS)

The difference between the narrow and wide graze is in the pattern the flock takes while foraging. The narrow graze is more military in style—the flock stays in formation while its members forage. In contrast, sheep foraging in the wide graze form a generalized, free flowing pattern.

The general deductions for the wide graze and road work apply to the narrow graze as well. For example, judges deduct points on the wide graze for the way in which a dog enters and exits the grazing area, protects the boundary, and runs in the furrow. Points deducted in the roadwork exercise are for failure to traverse the flock, causing splits, not working the opposite side and incorrect side changes.

As the handler enters the narrow graze area, the dog should move to a side of the graze opposite the handler. Failure to do so will cost the team up to 3 points. The narrow graze requires teamwork. The dog and handler work opposite sides of the flock, each controlling his own side.

When a handler is not using a second dog, he may use his shovel to toss dirt at sheep who do not keep within the boundaries. Otherwise, he should use his second dog to control his side of the flock.

If handlers fail to keep their side of the flock under control (regardless of whether or not they use a second dog) their team will lose 2-3 points. The handler should walk along side of the last third of the flock while the sheep graze. Failure to do so will cost 2-3 points.

If the sheep do not graze handlers may move to the head of the flock to lead them further into the grazing area and find better forage. Handlers may also need to place the dog before the flock if the sheep run ahead, not stopping to graze. Most judges want to see dogs change sides on command at least once during this exercise and will deduct one point if they fail to do so. Failing to execute a smooth change of sides may cost the dog and handler an additional 2 points. The turn into the graze to change sides must be at a right angle or the team will lose up to 2 points. Once the dog has changed sides he should move to the tail of the flock. Remaining at the head of the flock for too long a time can cost the team up to 2 points.

Photograph 11.13 Once in the narrow graze, the man dog works the handler's side of the flock while the main dog works the opposite side.

Chapter Eleven: Competitions

Photograph 11.14 After entering the narrow graze, the handler and dog stand as sentries.

Photograph 11.15 Change of sides during the narrow graze. This shepherd chooses to move forward to the flock's head as his dogs changed sides.

REPENNING (SV 4 Points)

As the flock nears the pen, the sheep may increase their speed. They know they are safe there and rest from the day's journey is just ahead, but shepherds must control their flocks and keep them from moving too quickly. If the flock is pressing onward too fast and the shepherd has only one dog, he may need to bring that dog from a position along side of the flock to the front. This is a judgment call on the handler's part. Once the shepherd reaches the pen he should take the position held by his dog while the flock exited the pen, i.e., along the latch side of the gate. He then directs his dog to take a position opposite him. From these two positions a shepherd and dog can oversee the flock as the sheep return to the pen. The shepherd can identify health problems and count their charges. If some sheep refuse to enter the pen, the dog should press them into it. This should be done calmly and quietly.

Once the flock is in the pen, dogs should position themselves in front of the opening. This allows handlers to secure the pen. The following two deductions are worth one point each: the first is when the position of the dog is too far or too close to the gate, thus splitting the flock; the second is when either the dog or handler does not stand at the correct position. If the flock moves around the pen instead of into it, judges can deduct up to 2 points.

Photograph 11.16 Dog and shepherd stand opposite each other as the flock enters the pen.

Photograph 11.17 Dogs protecting the entrance/exit of the pen as their handler prepares to close the gate.

GRIP (SV 8 Points)

Every flock has its stubborn members, those who show little respect for a dog. To contain and move the flock from place to place, the dog must have the flock's respect. To gain it, dogs may have to physically correct their charges from time to time. This must be done with a quick grip administered with the proper pressure. The permitted locations for grips are the neck, thigh, and rib. Generally dogs prefer to grip at a particular location, although some grip at all three. Dogs must not tear or shake their head while gripping.

Handlers signal their dogs either to grip or stop gripping. In either case the dog must respond immediately. At the beginning of a run, handlers should tell the judge when their dog will grip. If a dog refuses to grip on command or has not been trained to do so, judges may deduct 2-3 points for failure to be a useful tool.

A well-conditioned flock requires little reminder regarding the consequences of disobedience. However, if a grip is warranted but not administered at an acceptable location, the team could lose 2 or even more points. If a dog grips unnecessarily, judges will deduct 3-4 points.

In Germany judges will ask handlers to give a grip command on the way to the pen if their dog has not shown a grip during the course of the run.

Photograph 11.18 Dutch Shepherd rib gripping a sheep on command.

INDEPENDENCE (SV 10 Points)

When the dog performs his tasks with a minimum of commands, he demonstrates self reliance. Good tending dogs take responsibility for overseeing their flock. They watch their sheep, keeping them within the proper boundaries either on the road or while they are grazing. Such dogs also take responsibility for overseeing the flock throughout the day. If a dog reacts too slowly when members of the flock break away, judges may deduct up to 2 points.

Judges also want to see initiative. If they do not, a severe deduction may result—5 or more points. But if a handler sees a situation developing and commands his dog before the dog can react, judges deduct fewer points—up to 2.

OBEDIENCE (SV 10 Points)

Obedience is the foundation of dog training. To herd livestock dogs must be willing to accept guidance, responding eagerly and quickly to their handlers' commands. Their demeanor should reflect joyful submission to their masters' will. They should never demonstrate unwillingness to obey. If a dog must be asked more than once to take a command or if he executes a command grudgingly, judges will deduct up to 2 points. Teams will receive deductions if dogs do not respond to commands, are unsure or afraid of situations, need correction, respond too slowly to commands, or need handlers to repeat commands. In each case judges may deduct up to 2 points.

DILIGENCE (SV 10 Points)

In this part of the judging, dogs must show they are hard-working defenders of their flock and crops. Judges will deduct severely if a dog is lazy. For example, they will deduct 3 points if the dog does not ward off with speed. In addition, dogs should know where their sheep are at all times. If a dog must be reminded of this, judges will deduct 2-3 points. The dog's demonstration of good temperament and steadfastness of nerves is also important. The dog that does not show these qualities can lose 1-2 points and sometimes even more.

PROTECTION WORK

Dogs must be calm, controlled and have even temperaments. At the same time, they should always defend both their flock and handlers when necessary. Once a threatening situation is under control, however, their demeanor should once again be calm and controlled. Judges will critically score dogs that do not rise to defend their flocks or their masters.

Protection work is generally tested at the completion of the herding section of the trial. Dogs are placed in a situation that demands they come forward in defense of their masters. Generally an assistant is called on to engage the defense drive of the dog. The assistant-helper[35] will approach the dog in a threatening posture. The dog should

---

35   Generally a trained Shutzhund helper is used for this test.

notice this behavior and intently watch him. If the intruder continues to move toward the dog, he should bark, warning the intruder to come no closer. In some cases the intruder will not stop and the dog is then allowed to bite the helper's padded sleeve. When the intruder is no longer a threat, the dog must immediately become calm. The protection section of the trail is done with the dog on lead and is scored pass or fail.

Photograph 11.19 German Shepherd Dog alerting to an intruder's presence.

Photograph 11.20 German Shepherd Dog gripping the sleeve of an intruder.

French Trials (Société Centrale Canine)

The French competition comprises five exercises: pen, difficult passage, stopping the flock, conduct and maneuver, and intelligence of execution. It is generally 1,000 to 1,200 meters in length and is run in 25-30 minutes. French trials recognize three levels of competence. Level I dogs are scored as follows: 25 points for Pen, 20 points for conduct and maneuver, and 25 points for intelligence of execution, for a total of 75 points. No working certificate can be earned at this level. Level II dogs are scored 25 points for Pen, 15 points for difficult passage, 20 points for conduct and maneuver, 10 points for stopping the flock, and 30 points for intelligence of execution, for a total of 100 points. Dogs at this level work toward a title called a working certificate. They must earn a minimum score of 75% of the total score with at least 50% earned in each section.

Once dogs earn a Level II certificate, they may enter Level III competition. Dogs at this level are scored as follows: 25 points for pen, 25 points for difficult passage, 50 points for conduct and maneuver, 20 points for stopping the flock, and 30 points for intelligence of execution, for a total of 150 points. Teams can earn Level III Certificates in two ways: First, by placing 1st or 2nd with 80% of the total score, and second, by receiving three passing scores with a "very good"[36] rating given by different judges. Only level III dogs are allowed to compete in the French National Trial.

Pen (25 points)

The pen exercise is divided into three sections; exit, re-pen and protection and jump. The protection and jump exercise may be done either at the beginning of the pen work or after re-penning the livestock. Doing this at the beginning, however, gives handlers a chance to closely evaluate the flock. It also allows dogs to introduce themselves to the sheep, thus making it clear from the beginning that you are in charge. At the same time, it gives handlers a chance to reconfigure the flock for a smooth exit from the pen.

To begin, the handler enters the pen with the dog. He clears a passage for his dog and gives the stock room to walk about the pen which simulates feeding. This can be a very dangerous time because the sheep may push against shepherds' legs if they enter the feeding troth area with grain. If dogs do not keep the flock away from their handlers, judges may deduct 2-5 points.

Then the handler exits, leaving the dog to guard the open pen. If the dog leaves the open gate unattended judges can deduct 2 points for anticipating the departure of the flock. If the sheep leave the pen judges may deduct 2-8 additional points because

---

36 "Excellent" rating – 75% and above, "Very Good" rating—60% to 74%, "good" rating – 50% to 59% and "Non qualifying" – less than 50% of total points available.

the dog failed to control the sheep, thus allowing them to escape. Judges may also deduct 2-5 additional points if the sheep run away from the pen for a distance more than 50 meters. Moreover, if the sheep leave the pen while the dog is standing within the opening judges may deduct 2-8 points.

Once a dog shows he can hold the flock in place, the handler closes the gate with the dog still in the pen. The dog is then recalled to the handler. The dog must jump over the fence (one meter high) and return to its master's feet. If a dog does not do so, judges may deduct 2-5 points.

Dogs may enter the pen to exit the flock by either jumping in or going through the gate opened by the handler. After entering the pen, dogs should immediately take control of the stock. The flock should then exit the pen in a smooth unhurried manner. If a dog exits the flock too slowly or too abruptly judges can penalize the team 2-5 points. Once the flock exits the pen the sheep should not circle back and re-enter the pen. If this occurs, judges can subtract 2-5 points from the team's score.

When re-penning the flock at the end of the course, the dog should stop and hold the sheep in place while the handler opens the gate. The re-entry should be as smooth and calm as the take from the pen. The same deductions apply.

Difficult Passages (15 or 25 points)

Difficult passages may include bridges, sorting pens, narrow roadways, and narrow passages between crops and hedges. When negotiating an artificial bridge both handlers and dogs must cross the bridge as if it were a natural obstacle. When passing livestock through a sorting pen, handlers may not enter. However, dogs may enter the pen to keep the sheep smoothly moving through the system. In all other cases handlers may move about as they deem necessary. If during the course of the competition they miss an obstacle, they may not go back and retry it.

When approaching a difficult passage, the livestock should move calmly and smoothly directly at the obstacle. The flock should not move in a hurried manner and should not zigzag about. Doing so costs 2-5 points. If a flock goes around an obstacle (the bridge, for example) and does not pass over it judges may deduct 2-5 points. If, in the judge's opinion, handlers do not place their dogs correctly to help the flock negotiate an obstacle, the team can loose 2-5 points. If a flock leaves a narrow roadway and ventures into a bordering field, the team can loose up to 5 points. If a sheep becomes stuck in a sorting chute because of a dog's action and the handler fixes the problem himself the judge may deduct 2-5 points.

Judges may also deduct 2-5 points if a flock negotiates the obstacle late, negotiates it poorly, or looses control of the livestock as they exit the obstacle. Suppose the flock stalls at the mouth of an obstacle for an extended period of time, but eventually negotiates it. They have negotiated the obstacle but negotiated it late. Or suppose the flock is in the sorting chute but the execution is poor. For example, suppose the dog allows members of the flock to move forward but some get stuck and others jump out of the chute. This constitutes poor execution.

Management and Maneuver (20 or 50 points)

The Management and Maneuver section comprises five exercises: graze, holding and catching a sheep, distance work, traffic, and movement. The grazing area is either square or rectangular. Its boundaries are marked by four stakes, one in each corner. The area measures either 12 by 12 meters (for flocks of 50-70 head) or 15 by 15 meters for larger flocks.

Once the flock is settled in the graze, the dog and handler team position themselves at opposite corners. They wait there for a signal from the judge to proceed, all the while keeping the flock within the marked area. Dogs should calmly and efficiently contain their flocks in the designated area on their own initiative. If a dog has difficulty immobilizing the flock judges may subtract up to 5 points. If a handler must command his dog to contain the flock during the graze judges may deduct up to 1 point. If a dog is not in the necessary location to contain the flock while the sheep graze, judges may subtract up to 1 point. If a dog puts the flock back in place but does so late, judges may deduct as many as 5 points.

Once the flock is contained in the grazing area, handlers must catch and hold a marked sheep. If a handler appears unsure while catching or holding the animal, the team may lose up to 2 points. In some instances judges may ask handlers to clip a hoof or milk a sheep. While the handler is concentrating on this single sheep, the dog should contain the rest of the flock in the grazing area.

Handlers must then ask their dog to work at a distance (150-300 meters away). They have two options. First they may allow their dog to contain the flock in its specified grazing area while they go to a location designated by the judge. If during this time a member or the flock leaves the grazing area, judges may deduct 2-5 points. Similarly, if a dog leaves the flock and returns to its handler, judges may subtract as many as 5 points from the team's score.

Alternatively, a handler may go to the specified location accompanied by his dog. The judge will then tell the handler when to send his dog to gather the flock from the grazing area. In gathering the sheep from the grazing area the dog must move the flock along a designated path.

Dog-handler teams must then take their flocks along a road while a vehicle passes by. The road must be clearly marked. If simulated, a roadway must have at least one side free,[37] allowing room for a vehicle to pass by the flock. If a vehicle has difficulty doing so judges may subtract 2-8 points. If the team does not immobilize the flock while the vehicle passes, judges may deduct up to 5 points.

---

37 "Excellent" rating – 75% and above, "Very Good" rating – 60% to 74%, "good" rating – 50% to 59% and "Non qualifying" – less than 50% of total points available.

Finally, flocks should move through the course between exercises smoothly and efficiently. Teams should keep their flocks together and move sheep in straight lines from point to point. If a flock does not do so but instead winds its way through the course or moves in a choppy manner, the team can lose as many as 8 points. If the flock moves too slowly or too quickly, judges may deduct up to 5 points.

If at any time during this section of the course a flock escapes from a dog judges can deduct as many as 10 points. Moreover, if a dog scatters or jostles his sheep 2-8 points may be lost. If the handler and dog loose control of the sheep and the flock shows little progress moving through the course, judges may deduct 2-10 points.

Stops (10 or 20 points)

Dogs should be able to stop their flock at any time. If they try to do so but fail, judges may deduct as many as 10 points. If a handler helps stop the flock, the team may lose 2-8 points. The dog must come to the front of the flock to stop its forward motion. If a dog continues working the flock ignoring its handler's command to come to the front, judges may subtract 2-8 points.

Once the dog stops the sheep, the handler should determine whether or not the flock can safely continue forward (especially at a roadway intersection). Handlers should look for on-coming traffic and only proceed forward when they see none.

This pause is not long in duration. When the flock can safely continue, handlers should command their dogs either to drive the sheep forward or control the flock's sides. If a flock does not stop at the designated point, the team will loose all available points. If one sheep does not stop but passes the designated point instead, the team will lose up to 5 points. If, however, the dog fails to control the rest of the flock, judges will deduct 2-5 points. If the flock begins to reverse its direction after it stopped, the team may lose 1/2 of a point. If a dog has difficulty restarting the forward movement of the flock, judges may deduct 2-5 points.

Intelligence of Execution (30 points)

This section is judged throughout the competition and is divided into three parts: commands, obedience and activity. Handlers may give commands or cues either by voice, by whistle or by gesture (hand or staff). In either case, all commands must be consistent (the same cue for the same action) and used sparingly. Judges may deduct up to 3 points for useless commands or commands not executed by a dog. If handlers give their dogs commands with too much authority, the team can lose up to 3 points. If a dog loses interest in working the flock and his handler must recall him to resume work, judges can deduct up to 3 points. If handlers must correct their dogs, the team can lose as many as 6 points. When a command is given that does not make sense to the judges, they can deduct 2-5 points.

Obedience to a command should be immediate and commands should be carried out without question. When a dog does not obey a command, judges may deduct up to 1 point. If a dog is slow in obeying commands, judges may subtract up to 3 points. If a dog obeys "counter to sense," judges may deduct 2-5 points. For example, when a handler commands a dog to move to the left and he moves to the right instead, this constitutes a command obeyed "counter to sense:" Any time a dog abandons his flock and refuses to work, the team is disqualified.

Dogs should work with gentleness and initiative throughout the course. They should guide, stop and control the flock at their handler's behest. Yet, they should also work on their own initiative, returning strays and holding the flock in place.

When a dog lacks interest in the work at hand judges may subtract 2-5 points. If a dog allows members of the flock to escape the team may lose up to 3 points. When a dog becomes focused on a particular sheep or position and does not respond to the handler's command to move, judges may deduct up to 3 points.

Dogs should keep their flocks grouped together and not crash through them causing sheep to split or scatter. Dogs that do so may cost a team up to 3 points. Dogs that are fearful of situations or who do not assert themselves when challenged may cost their team up to 3 points. Dog who wander or chase members of a flock can cost their team up to 3 points.

In general, dogs should not bite the sheep. But the rules permit a nip to reinforce dominance. Brutalizing a sheep by roughly gripping it, however, will cost 2-5 points. Unnecessary, harmless grips may cost as many as 5 points. But judges will not permit unnecessary, flagrant grips and will disqualify a team for such an infraction.

Disqualification results from the following:
1. A dog straying on the course while another is competing (the dog who strayed on the course is disqualified);
2. An inability to control the livestock;
3. Aggressiveness;
4. Frequently scattering the flock;
5. A dog or handler repeatedly brutalizing livestock;
6. Harsh behavior by handlers toward their dogs;
7. Unjustified vicious grips;
8. Flagrant disobedience;
9. Abandoning the flock;
10. Lack of progress;
11. An unjustified dispute.

Tips

The French and German courses require some slightly different skills in your dog. First, the grazing exercise in the French course is different from the German course. In the French course flags or cones mark the boundaries of the grazing area. Therefore, your dog must look for these markers instead of feeling the boundary with his feet and legs.

One way to help your dog find these markers is to use a go to an object command (see page 179). Upon receiving the command your dog will begin to look for a marker to go to. Once there, give him another go to an object command until he has found all four sides of the grazing area. When your dog is familiar with the boundaries, he can learn to patrol between markers.

The second difference is that the French course allows an outrun, lift and fetch—in other words a gather. These elements are similar to the placement before the flock exercise. The difference is that you have no boundaries to follow.

Once in the grazing area the sheep are left grazing while the handler proceeds to the handler's post 250 yards away, as we saw earlier. Handlers have the option of taking their dogs with them at this time or leaving them with the flock and then recalling the dog once the handler is at the post. After the dog/handler team is reunited at the post and on the judge's signal handlers command their dogs to gather the flock, moving the sheep to the handler's post. Training your dog to gather is described in the following books:

1. *Lessons From A Stock Dog* by Bruce Fogt;
2. *A Way of Life* by Glyn Jones;
3. *Herding Dogs Progressive Training* by Vergil Holland;
4. *Farmer's Dog* by John Holmes;
5. *Anybody Can Do It* by Pope Robertson;
6. *Sheepdog Training, All Breed Approach* by Marie Taggart;
7. *The Traveling Herding Teacher* by Bob Vest.

A third difference between the French and German course relates to biting. That is, a grip verses a nip. Check with your judge to find out which type of bite he/she will permit. In German competitions the rules permit a full mouth grip at the neck, ribs or upper thigh. Judges may not allow these grips in French trials. They will, however, allow nips at the heels or noses of animals. Therefore, you will probably need to teach your dog to bite on command in a different manner.

ANKC, CKC and AKC Trials

Both Australian National Kennel Council and Canadian Kennel Club herding competitions mirror the American Kennel Club herding program. The American Kennel Club "C" course is a combination of the German and French tending courses. It resembles the German course in some ways but with French "Stops", time limits and three levels of competition added. The maximum number of available points in each class is 100. The Started class includes an exit and re-pen (15 points), narrow road (20 points), bridge (10 points), pause-traffic (20 points) and wide graze (35 points). The course is run with a 30-minute time limit along a course 440 to 540 yards in length.

The intermediate class includes an exit and re-pen (15 points), narrow road (20 points), bridge (10 points), pause/traffic (20 points) wide graze (25 points), and placement before the flock (10 points). The time limit is 45 minutes; the course is 540-780 yards in length.

The advanced class includes exit and re-pen, (15 points), narrow road, (20 points), bridge, (10 points), pause/traffic, (20 points), wide and narrow grazes, (25 points) and placement before the flock, (10 points). The time limit for this course is also 45 minutes, but the course length is 780 to 880 yards.

Because tending competitions are new to herding enthusiasts in the North American, the judging of this course tends to vary and the rules change. Some judges make decisions about scores that resemble those made in German trials, while others score more according to French rules. Competitors should study the current rules carefully and ask questions of judges before competition begins.[38]

FCI (Fédération Cynologique Internationale) Traditional Style Herding Competitions

The FCI holds both working tests and trials throughout Europe. They divide these tests and trails into traditional and collecting style courses. The collecting style course is open to Border Collies and Kelpies only. Traditional style competitions are open to any cattle or sheep dog listed as such (FCI Groups 1, 2, and 5) excluding Border Collies and Kelpies. Although both sheep and cattle dogs are eligible to enter working tests or trials, the only livestock used is sheep. For the working test 10 sheep are used. For trials 15-80 sheep are used. Larger flocks are preferred "because they give a better opportunity for success" (International Rules for Sheepdog Trials—Traditional Style page 241).

The FCI traditional style working test consists of the following exercises: pen (20 points), grazing (20 points), conduct and maneuver (30 points), behavior to sheep, gentleness (20 points) and obedience (10 points). A minimum of 60 out of 100 points is needed to qualify. The following is taken from the FCI traditional style test regulations.

Behavior Test

It is up to the judge to decide how to test. The dog should be tested as to its natural behavior. It is not allowed to attack or to stress the sheep. It can be done while the handler is leading his dog (on leash) through a small group of persons while the judge is talking to him. During this test the dog's tattoo number or micro-chip is verified.

The dog must be natural and lively. Shy or aggressive dogs are not allowed to run. A little reserve to strangers is accepted.

---

[38] To obtain a current copy of the AKC Herding Regulations, Contact: The American Kennel Club, 5580 Centerview Drive Suite 200, Raleigh, NC 27606-3390, (919) 233-9767 (website – www.akc.org).

Course Description

Pen or Sheepfold

Exit: On the judge's command the handler may open the pen for the dog to enter (he can accompany the dog or not), to accomplish the exit of the flock. The exit is accomplished by the calm and active work of the dog under the command of the handler. The dog should immediately take control of the stock. The judge evaluates the behavior of the dog in contact with the flock—its calmness and firmness.

Re-entry: Re-entry should be accomplished by the active work of the dog under the command of the handler: It must be calm and the sheep should not be jostled. Near the sheepfold, the flock should be halted and held in place by the dog, while the shepherd opens the gate.

Conduct and maneuver:
1. After the exit from the pen the flock must be moved to the grazing area. The distance from the pen to this area is at least 75 m while the handler is using a simulated road or path. Judges evaluate both the movement of the flock throughout the test and the dog's ability to guide the flock through the course.
2. Graze: The work consists of grazing to the square or elongated in a natural field or temporary field with its limits marked by four stakes at the corners or along the front of an authorized space (15 x 15 m). The judge appraises the shepherd's handling, the position of the dog in relation to the flock and the ability of the dog to bring the flock into the graze and keep it in place with calm, efficiency and initiative.
3. Stop (immobilization): the dog should be capable of stopping the flock. This can be done when the handler comes back from the grazing area and moves to the pen for re-entry. The judge indicates the point at which the flock should be stopped.

Intelligence of execution

1. Commands: Commands can be done by voice, gesture or whistle and should significantly influence the behavior of the dog and the quality of its response. Judges look carefully at how the dog responds to these commands.
2. Obedience: This should be immediate, complete and definitive, for it conditions good control of the flock.
3. Activity-initiative-gentleness (behavior towards the sheep): The judge evaluates the ability of the dog to guide, stop or move the flock according to the handler's instructions, as well as the ability of the dog to intervene on its own initiative and upon instruction, direct, stop or hold the flock in place, and regroup strays.

Apart from exceptional cases, the dog should not bite the sheep and all brutality is severely penalized. A brief nip is admissible, only when necessary and in order to keep control of the flock. Obvious and untimely bites are the cause for immediate disqualification and the dog should not endanger the sheep.

Point deductions: (-)

1. Pen or sheepfold
   - anticipates departure (2).
   - too hurried or too slow in work (2 – 5).
   - allows re-entry (2 – 5).
   - allows runaway of more than 30 meters (2 – 5).
   - does not work, does not control sheep (2 – 8).
   - does not keep the sheep back from handler when both are inside pen (2 – 5).
   - does not jump (2 – 5).
   - allows runaway of more than 50 meters (2 – 5).
2. Conduct and maneuver
   - dog poorly placed (1).
   - goes through the course too rapidly/too slowly (1 – 5) .
   - late in putting the flock in place (1 – 5).
   - difficulty in immobilizing the flock (1 – 5).
   - tentativeness in catching/holding the marked sheep (2).
   - choppy, winding, imprecise transit of the course (1 – 8).
   - scattering or jostling the sheep (2 – 8.
   - grazing out of limits (2 – 5).
   - loss of control, little work (2 -10).
   - does not immobilize sheep (5).
   - movement of flock during "hold" and "stops" (5).
   - dog returns to handler during graze (5).
   - complete escape of flock (10).
3. Stop
   - a sheep that pushes past (1 – 5).
   - sheep pull back too far (0.5).
   - slight overflow from the flock, but controlled (2 – 5).
   - late in resuming motion (2 – 5).
   - stop done by the handler, not the dog (2 – 8).
   - dog keeps working, does not stop (2 – 8).
   - dog allows the entire flock to move – the flock does not remain at a complete stop (10).

4. Intelligence of execution
    a. commands
        - useless or not performed (1 – 3).
        - too numerous, lacking in firmness (1 – 3).
        - wrong command (2 – 5).
        - recalling the dog on its abandoning work (3).
        - correcting the dog (6).
    b. obedience
        - ignores the command (1).
        - late in obeying (1 – 3).
        - obeys command in a wrong way (2 – 5).
        - abandons the flock (disqualification).
    c. activity - initiative - gentleness
        - wandering, chasing (1 – 3).
        - difficulty in asserting itself, fearfulness cutting into the flock, scattering the flock (3).
        - lack of interest in work, inattentive (2 – 5).
        - dog places itself poorly, allows escapes (1 – 3).
        - does not move when ordered (1 – 3).
        - brutality, rough gripping (2 – 5).
        - unnecessary grip (1 – 5).
5. Disqualifications
    a. Handler Disqualifications.
        - drunk and/or under the effect of drugs.
        - unjustified dispute.
        - allowing dog to wander during another competitor's run.
        - brutality toward sheep or dog.
6. Dog Disqualifications
    - disrupting the normal sequence of the course.
    - abandoning the flock.
    - refusal to obey, flagrant disobedience.
    - unjustified or dangerous bites.
    - repeated brutality of the sheep.
    - frequent scattering of the flock.
    - fearfulness or aggressiveness.
    - failure to control the flock.

Herding Trials – Traditional Style

FCI traditional herding trials were approved in 2007 and are similar to French herding competitions. The difference between the French course and the FCI course is that the French course includes a gather described in the French course rules as follows.

"Handlers must…ask their dog to work at a distance (150-300 meters away). They have two options. First they may allow their dog to contain the flock in its specified grazing area while they go to location designated by the judge. Similarly, if during this time a member or the flock leaves the grazing area, judges may deduct 2-5 points. Similarly, if a dog leaves the flock and returns to its handler, judges may subtract as many as 5 points from the team's score.

Alternatively, a handler may go to the specified location accompanied by his dog. The judge will then tell the handler when to send his dog to gather the flock from the grazing area. In gathering the sheep from the grazing area the dog must move the flock along a designated path."

The FCI traditional trial course is meant to test the ability of traditional herding dogs rather than dogs like border collies and kelpies, and thus does not include a gather. The elements of the course depict everyday work including pen, leading, grazing, traffic and stops. According to the FCI, "the aims of … (traditional) sheepdog trials (is) to promote good handling of the sheep with a minimum of stress for them as well a good handling of the sheepdog, to promote its utility and sport, and to secure and select the best working lines". Truly both form and function are important elements of FCI traditional herding competitions.

FCI traditional herding competitions offer three levels of difficulty. A dog must earn a score of 70% on a working test to qualify for entering all FCI herding trials. After earning a score of 70% or better in a trial level class a handler has the option of moving to the next higher level. Once a dog has received three scores of 70% or better the dog must move to the next higher level.

Class 1
75 points (about 25 minutes)
1. Pen or sheepfold (25 points).
2. Difficult passages (none required).
3. Conduct/Maneuver (20 points).
   Graze, hold, catch a sheep.
4. Stops (none required).
5. Intelligence of execution (30 points)

Class 2
100 points (about 30 minutes)
1. Pen or sheepfold (25 points).
2. Difficult passages (1 or 2 + bridge) (15 points).
3. Conduct/Maneuver Graze, hold, catch a sheep (20 points).
4. Stop—one stop (10 points).
5. Intelligence of execution (30 points).

Class 3
150 points (about 35 minutes)
1. Pen or sheepfold (25 points).
2. Difficult passages (2 or 3 + bridge - 25 points).
3. Conduct/Maneuver Graze, hold, catch a sheep-Traffic (50 points).
4. Stops (two stops - 20 points).
5. Intelligence of execution (30 points).

Exercises

A. Pen or sheepfold.

A jury of judges (a single judge in less important trials) evaluates dogs according to how skillfully they perform two required exercises – exiting the stock from the pen and re-entering the pen. In FCI traditional style trials protecting the handler and jumping out of the pen after the sheep have exited are considered part of these two exercises.

Exit: After the authorization from the judge, the handler may open the pen for the dog to enter. The handler may or may not accompany the dog to accomplish the exit of the flock. Regardless, the exit is accomplished by the active work of the dog obeying the commands of the handler. It should be calm, and the flock should take control of the flock as immediately as possible. Jury evaluates how the dog takes control of the flock—it should do so calmly and firmly.

Re-entry: Re-entry should be accomplished by the active work of the dog under the commands of the handler. It should be calm and the sheep should not be jostled. Near the sheepfold the flock should be halted and held in place by the dog while the shepherd opens the gate.

Protection and Jump: Before the exit of the flock from the enclosure where it is contained or at the time that it has re-entered, the gate is closed. The handler simulates feeding and going around the paddock. The dog clears a passage between the shepherd and the flock in order to protect shepherd's entrance or exit from the pen. Either at the beginning or end of the exercise the dog is placed between the flock and the gate in order to protect the handler's entrance or exit from the pen. The handler leaves through the gate, leaving the dog protecting it. After the gate is closed the handler recalls his dog, which must jump over a panel in order to return to his place at the feet of the handler.

B. Difficult passages.

There should be two (or even three) difficult passages, which should mimic the real world as much as possible, for example, a narrow passage between crops, hedges, barriers between two fields, or a passage over a bridge or sorting pen.

C. Conduct and maneuver.

Graze: The work consists of the graze, either square or elongated in a natural field or temporary field with its limits marked by four stakes at the corners or along the front of an authorized space (15 x 15 meters).

The jury evaluates the shepherd's handling, the position of the dog in relation to the flock, and the ability of the dog to bring the flock into the graze and maintain it in place with calmness, efficiency and initiative (time is up to the jury's discretion).

While the flock is immobilized and calm in the authorized space, the handler must catch and briefly hold a sheep marked by the jury. During this time the dog should contain the rest of the flock in the authorized space – on its own initiative, if possible.

Car passage: This may be done on a real road or on a simulated road that represents the conditions encountered on a real road. Simulated roads must be clearly delineated, and wide enough so that the dog can press the flock to one side, leaving enough room for a car to pass on the other.

Movement: The jury judges the movement of the flock throughout the duration of the course. The dog will be judged for his ability to guide the flock between course exercises or obstacles, taking into account the relative difficulty and number of these movements.

D. Stops.

The dog should be capable of stopping the flock in all circumstances. There should be at least two stops, placed at natural places—paths, roads, enclosures, fields, etc. The jury distributes the points for this part of the competition based on the relative difficulty of the stops.

Stopping the flock is a very important maneuver and should be accomplished by the dog coming to the front of the flock. After the flock stops and the handler checks that the passage is not dangerous, the flock's resumption of movement should be fairly quick. The sheep should not be jostled and the dog should go around the flock to push at the rear or at least assure the control of a side. Going past the stop results in the loss of all the points for this exercise; the handler cannot attempt a new stop.

E. Intelligence of execution

This is judged throughout the whole trial and divided into four headings.

1. Commands: This may be made by voice, gesture or whistle and should influence the behavior of the dog and the quality of its response. The jury evaluates the vocabulary, clarity, force, intonation and number of commands. Useless commands or commands not followed by the dog are penalized. Each sound of the whistle is a command. Whatever the mode, the commands applied to each movement should always be the same and used as sparingly as possible.
2. Obedience: Obedience should be immediate and complete, for it conditions good control of the flock.
3. Activity (initiative, gentleness, behavior): The jury evaluates the ability of the dog to gather, guide, stop or move the flock both with a minimum of force and his ability to intervene on his own initiative and channel, stop or hold the flock in place and regroup strays.
4. Behavior to Strangers: The behavior of the dog toward strange people is noted. It should be natural—lively, but neither shy nor aggressive. Some reserve is acceptable.

Except in special cases the dog should not bite the sheep; all brutality is severely penalized. A brief nip is permissible, but only when necessary to move a recalcitrant sheep and to keep control of the flock. Flagrant and untimely bites result in immediate disqualification.

Point deductions: (-)

A. Pen or sheepfold
- anticipates departure (2).
- too hurried or too slow in work (2 – 5).
- allows re-entry (2 to 5).
- allows runaway of more than 30 meters (2 – 5).
- does not work, does not control sheep (2 – 8).
- does not keep the sheep back from handler when both are inside pen (2 – 5).
- does not jump (2 – 5).
- allows runaway of more than 50 meters (2 – 5).

B. Difficult passages
   - sheep step out of limits (0.5 – 5).
   - poor approach (2 – 5).
   - dog poorly placed (2 – 5).
   - late in negotiating of passage (2 – 5).
   - bad negotiation of passage (2 – 5).
   - intervention of handler (2 – 5).
   - loss of control when exiting passage (2 – 5).
   - flock goes around or does not pass through (2 – 5).
C. Conduct and maneuver
   - dog poorly placed (1).
   - goes through the course too rapidly/too slowly (1 – 5).
   - late in putting the flock in place (1 – 5).
   - difficulty in immobilizing the flock (1 – 5).
   - tentativeness in catching/holding the marked sheep (1 – 5).
   - choppy, winding, imprecise transit of the course (1 - 8).
   - scattering or jostling the sheep (2 – 8).
   - grazing out of limits (2 - 5).
   - dog poorly placed - car passage (2 – 8).
   - loss of control, little work (2 – 10).
   - does not immobilize sheep (5).
   - movement of flock during "hold" and "stops" (9).
   - dog returns to handler during graze (5).
   - complete escape of flock (10).
D. Stops
   - a sheep that pushes past (1 – 5).
   - sheep pulled back too far (0.5).
   - slight overflow from the flock, but controlled (2 – 5).
   - late in resuming motion (2 – 5).
   - stop done by the handler, not the dog (2 – 8).
   - dog keeps working, does not stop (2 – 8).
   - allowing total movement (10).

E. Intelligence of execution
   a. Commands
      - useless or not performed (1 – 3).
      - too numerous, lacking in firmness (1 – 3).
      - wrong command (2 – 5).
      - recalling the dog on its abandoning work (3).
      - correcting the dog (6).
   b. obedience
      - does not obey command (1).
      - late in obeying (1 – 3).
      - obeys command in a wrong way (2 – 5).
      - abandons the flock (disqualification).
   c. activity - initiative – gentleness
      - wandering, chasing (1 – 3).
      - difficulty in asserting itself, fearfulness (3).
      - cutting into the flock, scattering the flock (3).
      - lack of interest in work, inattentive (2 – 5).
      - dog places itself poorly, allows escapes (1 – 3).
      - doesn't move when ordered (1 – 3).
      - brutality, rough gripping (2 - 5).
      - unnecessary grip 1 (max. 5)
   d. Disqualifications
      1. Handler Disqualifications
         - drunk and/or under the effect of drugs.
         - unjustified dispute.
         - allowing dog to wander during another competitor's run.
         - brutality toward sheep or dog.
      2. Dog Disqualifications
         - disrupting the normal sequence of the course.
         - abandoning the flock.
         - refusal to obey, flagrant disobedience.
         - unjustified or dangerous bites.
         - repeated brutalities by the dog.
         - frequent scattering of the flock.
         - fearfulness or aggressiveness.

Twelve

Tending Applications

In preceding chapters you learned about the tending method of herding sheep and how to teach your dog to do it. In this final chapter, we focus on tending applications: how to use the tending equation in real life situations. In particular I will discuss some practical applications of tending for American farm operations.

Conventional wisdom says that no practical uses for tending on American farms and ranches exist. As one of the anonymous reviewers of an early version of this book said:

> I seriously doubt that from a practical standpoint there would be much use or interest (in tending for) the North American farmer or rancher. From my experience, good dogs can be extremely useful to livestock owners but not in the way boundary dogs are used.

In this instance conventional wisdom is both right and wrong. Boundary dogs (dogs trained in the tending methods described in this book) would not be very useful in most farming/ranching operations where sheep are raised solely for the money that their wool and meat fetches on the open market. In particular, tending offers little of value to large scale sheep operations in the Western Great Plains.

Unfortunately, many sheep operations, both large and small, are facing serious financial difficulties. In the pages that follow, I offer some new ways to use sheep profitably, ways that utilize parts of the tending equation. The key word here is "use." The practical applications of tending require us to rethink sheep-raising. We need to think of sheep as tools to use, as well as livestock to raise. Of course the applications I discuss are not a panacea for the sheep industry. Rather, they are suggestions that some sheep owners might find helpful in making their stock operation more profitable.

To begin, let's look at some of the ways livestock are used as tools in other parts of the world. Sheep have been used for centuries in Europe to cut grass, manage brush, and fertilize soil. For instance, in France sheep are placed in folds (large pens) to eat grass and fertilize the soil. Once this is accomplished, the fold is moved to a new location and the process begins again. Farmers either collect the manure and deposit it where needed (if their intent was to fertilize that location) or leave it where it is. In Germany, shepherds are paid by farmers and municipalities to graze their sheep on land that needs to be cut and fertilized. Often sheep will graze through an area after a crop is harvested in the fall, cleaning the field for a cover crop of winter rye. There are two benefits to this:

1. The owner of the land has his fields cut and fertilized;
2. The sheep owner feeds his sheep for free.

Finally, in Switzerland, sheep, goats and cattle are grazed in the Alps during the summer months. These sure-footed animals can easily graze the steep and dangerous alpine slopes where machines cannot go. When the herds and flocks are finished grazing, the pastures look as if they were professionally manicured.

Not all areas grazed by livestock look like they were mowed by machines, however. As we saw in earlier chapters, two methods of grazing exist: long and short term grazing. Traditional grazing methods keep livestock in areas for long periods of time. The livestock first grazes on the plants it likes best, leaving the less palatable ones. Then it often grazes on the more appetizing plants once again. This leaves areas where tall plants are interspersed among the grazed areas, giving the pasture a ragged appearance.

Short duration grazing or intensive rotational grazing on the other hand, concentrates the animals in "… smaller areas, forcing them to graze all plants evenly" ("Benefits" United States Department of Agriculture Soil Conservation Service, nd). This gives each pasture time to rejuvenate and, as we saw in an earlier chapter, allows the more desired or appetizing plants a chance to compete with the undesirable ones. Moreover, it reduces the amount of weeds and thickens the desired plant stand in the process.

As you will recall, the key to intense rotational grazing lies in taking the livestock out of the area before the animals begin to re-graze on new growth. Otherwise, they will kill the more palatable grass. When the pasture contains only palatable forage, the animals will cut the grass evenly, giving the area a manicured appearance. This system not only saves time and labor; it enables farmers to gain a longer grazing season.

Here in North America a few livestock producers have begun to use the intense rotational method of grazing, referring to themselves as grass farmers. This is because much of their profit comes from managing grass, using their animals as tools in the process. By making grass management a critical focus, they are able to reduce feed costs. In addition, land owners pay them to maintain their land, but just as important, the land gains environmentally friendly techniques and shepherds save labor.

In selected locations throughout North America, farmers and ranchers are successfully using these methods. True, tending is not yet the herding method of choice in North American grass management operations. Nevertheless, it offers particular advantages to shepherds who use their sheep as tools, as we shall see later in the chapter.

One example of a successful grass management operation comes from Western Canada. There, timber related work is a major source of income. After a forest is logged, seedlings are planted to replace the cut timber. Such logged-out areas are called cut blocks (or sometimes clear blocks). In such areas, unwanted vegetation must be kept low to allow seedlings sufficient sunlight to grow.

In the past, forest managers used three methods of weed control: mechanical, manual, and chemical. The mechanical method utilized heavy equipment. It was fast and therefore inexpensive, but not practical in locations with steep inclines and a correspondingly high potential for erosion, or where seedlings had been planted. Methods using hand-held cutting tools were effective but in some sites they were too expensive, costing $400 to $700 per acre. Herbicides were cost effective ($75 per acre) but were rarely used because of environmental considerations (Gerhard 1993:18). Consequently, timber companies found other methods of weed control.

One such method utilized sheep. Sheep were first used experimentally in 1984 in the Caribou region of British Columbia. Since then, sheep grazing has successfully been used as a means of vegetation management throughout the province (Schwantje 1995:1). Sheep owners are paid to graze their ecologically friendly sheep on weeds that otherwise would choke out seedlings ("For Hire: Sheep Labor," Success, June 1993:10). Although sheep are more expensive than herbicides ($5.00 a head per month) they are cheaper than mechanical and manual methods; contractors have seen a 25% reduction in costs using sheep. Moreover, they can also graze in areas where machinery cannot reach (Gerhard 1993:18-19), and when sheep graze in areas where seedlings have been planted, timber growth increases dramatically because of the manure.

Sheep Management in Forest Plantations

The following is adapted from *Interim Guidelines for the Use of Sheep for Vegetation Management in British Columbia*: 1995. I have included it to give you some idea of how shepherds work in cut block areas. It shows both the problems inherent in working this kind of environment and the solutions devised to deal with these problems.

Different sheep breeds have variable gregariousness or flocking rates which can affect flock management and grazing uniformity. For instance, the Rambouillet, Columbia and Corriedale breeds are more gregarious, making them easier to manage and thus allowing more uniform grazing on rough terrain.

In contrast, Dorsets, Suffolks, Hampshires and Cheviots tend to graze further apart, thus making them more difficult to manage. Yet, "it is recognized that some grazing situations are better managed by using flocks with a more dispersed grazing behavior" which requires increased supervision (Schwantje 1995:9).

Pens

Pen placement is very important. Sheep must be enclosed in pens every night. Pens should be located in non-productive areas, e.g., gravel pits or spur roads. They should also be located near roads and skid trails; this keeps the sheep from trampling productive areas. Additionally, pens or corrals should be:
1. dry and well drained;
2. free of debris;
3. large enough to pen the entire flock;
4. away from heavy brush or other predator cover;
5. free of poisonous plants;
6. at least 30 meters from water courses;
7. accessible to motor vehicles.

(Newsome, Wikeem and Sutherland 1995:11).

Pens should be portable and secure. They should be designed to provide effective service along with low maintenance. Therefore, they should: (1) be highly portable, making them easy to set up and take down; (2) require very few posts to be driven into the ground; (3) be manageable by two people; (4) be capable of withstanding hard use and require minimal repairs. Furthermore, their design must include a main holding area along with sorting and isolation pens for sick or injured animals (Schwantje 1995:11).

Nutrition

Sheep require 4.5 liters of water daily depending upon age and size of sheep, weather conditions and water content of forage. Sheep should be led to small pools of standing water. Additionally, they should be kept from contaminating continually flowing water sources and should be removed from water banks immediately after drinking to minimize erosion.

Sheep also require access to salt. The average sheep requires 5-14g of salt daily, again depending on size, age and the quality of forage. Salt should be located away from water sources but near resting areas (which should be rotated to avoid seedling damage).

Sheep grazing patterns can vary due to topography, weather conditions, illness and predation. Therefore, sheep must be given sufficient time to graze to insure adequate feed intake. Usually 8 hours per day is the minimum time (Schwantje 1995:9-12).

Herding

Government guidelines require both shepherds and herding dogs to be experienced. For instance, shepherds must understand silviculture,[39] sheep production, and the environmental objectives of cut blocks. Their dogs must also be skillful at moving and controlling sheep.

The ratio of shepherds and dogs to sheep is important. For a flock of 1,500 sheep, three shepherds should be on site with at least one herd dog per shepherd and one spare healthy herd dog available. For flocks under 1,000 sheep, two shepherds and two dogs are adequate. Guardian dogs should be used for additional protection against predators, with two dogs in a flock of 1,500 sheep (Newsome, Wikeem and Sutherland 1995:12).

Although two shepherds must stay with the flock at all times, one shepherd may leave, but for a period no longer than four hours. During this time the remaining, shepherd and dogs must stay with the flock (Schwantje 1995:12).

During the grazing season shepherds are responsible for the following:
1. checking the clear block for predators and wildlife each morning before releasing the flock to graze;
2. systematically grazing the block uniformly, making one pass[40] per grazing season and leaving 5-15% ground cover[41] (Newsome 1993:6);
3. maintaining flock health and individual sheep weight;
4. returning the flock to their night pen before dark and checking for strays;
5. staying with the flock at all times, which includes camping next to the pen overnight to discourage predators.

(Newsome, Wikeem and Sutherland 1995:12).

While grazing cut blocks, the shepherd's main responsibility is protecting the seedling trees. To do so, shepherds must move their flock slowly out of their night pens. When possible, they use skid trails and old roads to advance their flocks through the plantation of young trees. When traveling[42] daily between the night pens and the day's grazing area, shepherds must take alternate routes to protect the ground cover. When the sheep are grazing they should be allowed to spread out and graze independently. As they graze they should move slowly, constantly and systematically through the plantation (Newsome, Wikeem and Sutherland 1995:12).

Not only are the seedlings important to the shepherd, but the growth and performance of his flock are as well. Throughout the grazing season the sheep should have

---

39 Growing and tending to the growth of a forest.
40 At least two passes and some times three passes are required to effectively graze a plantation.
41 Studies show that seedlings can grow to 75 – 100 % of optimal growth performance with 5 – 15% competing vegetation cover (Newsome, Wikeem and Sutherland 1995:13 – 14)
42 In general a maximum of eight kilometers can be travelled per day on average terrain (Interim Guidelines for Use of Sheep.... 1995:12)

access to quality forage. When blocks do not produce enough forage, the flocks should be moved to another location or given supplemental feed. Generally, a clear block is not re-grazed unless sufficient re-growth exists.

When grazing a cut block shepherds must calculate the number of grazing days their flocks can forage in their assigned blocks. Generally, plantations are divided into smaller grazing areas whose maximum size is 50-60 ha.[43] Once shepherds make this division, they use the following calculation:

$$\text{Number of grazing days} = \frac{\text{Area (ha)} \times \text{Dry wt. of forage (kg/ha)} \times \text{\% Removal}}{2 \text{ kg/day} \times \text{Number of sheep}}$$

Where:
Area = area to be grazed in hectares,
Dry wt. of forage = oven-dried weight of target
Vegetation sampled on the site, % removal = amount of target vegetation removed by sheep to achieve the desired silviculture goals (Newsome, Wikeem and Sutherland 1995:14).

For example, if a shepherd has 1000 sheep, and a 50 ha clear block containing 1,000 kg per ha of available forage, he can graze his flock there for 12.5 days. This is only a rough estimate, however, and shepherds usually give themselves a buffer of at least 2 days. If the sheep finish grazing an area earlier than expected, the shepherds must have a back-up plan. Generally, they have two options: 1) leave the block early, moving to other back-up blocks; or, 2) increase grazing days by decreasing flock size (Newsome, Wikeem and Sutherland 1995:14).

Other Grazing Projects

Several other grazing projects in Canada also work well. The Parks and Recreation Department in Fort Saskatchewan, Alberta uses 200 sheep to graze 100 acres of public land on an abandoned provincial jail site ("Lunch on the Lawn" June 1992:6). The advantage of this is that fewer human employees are needed to cut the grass.

The town of Wetakawin, Alberta uses 200 sheep to cut the grass in a sewage lagoon. It costs the town approximately $4,000 Canadian to use sheep compared to $8,000 - $9,000 to mow the grass mechanically (Love 1994:25). Calgary Olympic Park also uses sheep successfully. Over the last few years park employees have used

---

[43] A hectare equals 2.47 acres

300-500 sheep to mow the grass in an environmentally friendly way to keep the lawns manicured (Love 1994:25).

Similarly, in the United States shepherds use sheep and goats successfully as grass and brush management tools. In the south-western United States researchers studying the effectiveness of using goats for brush control found that in one year 38% of the brush was eradicated in a test area of 56 acres, producing a savings of $840 in herbicide costs ("Using Goats for Brush Control in Humid Climates" 1991:7). More recently, Google and Yahoo in their ongoing efforts to go "green" are using goats rather than mechanical devices to keep grass trimmed on their corporate campuses (Miller: June 2009:46).

In the National Forests of the Pacific Northwest sheep producers use their animals to accomplish similar tasks. In that region sheep are an effective method of weed control because they do not prefer evergreen seedlings. Instead, they eat the unwanted weeds and other vegetation that compete with the newly planted seedlings ("Woolly Weed-eaters" 1992:10).

Due to the successful use of sheep in the Pacific Northwest where herbicides are restricted, researchers in Ohio are studying the use of sheep in weed control projects ("Sheep and Trees Mix Well" 1991:14). Soon we may see sheep grazing eastern state forests as well.

In fact, sheep are already used as vegetation management tools in Vermont and New Hampshire. In the spring of 1991, the town of Orwell placed an advertisement in the local newspaper. The town was seeking bids to mow the steep rocky hillsides of two cemeteries. Jean Beck answered the advertisement. She, like many of the professional landscaping companies, submitted their proposals to cut the grass in each cemetery. Ms. Beck's bid of $250 was roughly $3,000 lower than the next lowest bid.

How could she out bid the other contractors by so much? Her employees were 17 Dorset sheep ("Let the Sheep do the Shearing" 1991:14). Many felt that sheep grazing in a cemetery was disrespectful to the deceased, yet the sheep gave the cemeteries a well groomed appearance.

In 1998 the Public Service of New Hampshire (PSNH) and Bellwether Solutions (an environmental consulting firm) of Concord, New Hampshire began a pilot sheep grazing program on the state's high voltage transmission lines. The aim of the project was to ascertain the efficiency and economics of using sheep to graze the 1,800 miles of transmission lines.

The vegetation under the power lines grew over time and thus overgrew the power lines themselves. Each year about 360 miles of the overgrown lines were cleared using costly mechanical mowers, chainsaws and an excavator. Unfortunately, this process was noisy, besides disturbing to the terrain and wildlife.

A practical, environmentally friendly alternative was to graze sheep along the right of ways under the transmission lines. Unfortunately, New England sheep are generally barn fed, prima donnas wanting only the best fed to them on a silver platter. Thus, the project bought from Montana 500 Rambouillets, which were very good at foraging for themselves.

The flock arrived in April, 1998, each sheep in the flock weighing roughly 100 pounds. They were shorn, vaccinated, wormed and then introduced to the power line grazing areas. The sheep were used to graze the 400 foot wide target areas for approximately five months (May thorough September/October). During this time they ate the leaves from Maple, Birch, Cherry and Oak trees. As the sheep stripped the leaves and bark from the trees, the trees died.

Shepherds used portable electric fencing to contain the flock and a strip grazing system to control vegetation but "in some areas the sheep were allowed to graze for part of the day without fences…" (Broadbent 1998:17). The hired shepherd tended the flock with his two Border Collies and a Great Pyrenees. He used the Border Collies to control the flock while the Great Pyrenees protected it from predators. "On occasion the sheep needed to be moved longer distances and extra people were needed to help cross roads and bypass properties…" (Broadbent 1998:17).

Finally, sheep are used to graze ski slopes in the Green Mountains of Vermont. In 1992, David Major began the "Mountain Sheep Project." This project uses sheep to graze the ski slopes of Mount Snow, Killington and other Vermont ski areas. The sheep graze their way up the mountain slopes using intensive rotational grazing methods. They graze on the brush and grass of the steepest slopes where machine mowers cannot reach.

The advantages of this are two fold: first, ski areas can open the more challenging trails earlier because less snow is needed to cover them. Four feet of machine-made snow is needed on ungrazed areas compared to just a few inches on slopes grazed by sheep. Opening the slopes earlier attracts earlier season skiers, thus increasing the season's revenues (Daley 1994:16).

Tending Advantages

Until now, the tending method of shepherding has not been used in North American grass management. Still, it offers particular advantages in these kinds of situations because shepherds can easily graze sheep in areas without fencing. For example, on a Vermont ski slope Josh Moody typically works a ten hour day. The ideal day starts at 5:30 a.m. and involves visiting each group of sheep, feeding the guard dogs, checking the fences, and moving them when necessary. This is hard, tiring work and in Josh's own words, "Mistakes happen when your angry tired, frustrated, or in a hurry" (Daley 1994:16).

So how would a tending shepherd organize the work in the Vermont Mountain Sheep Project? First, sheep would be penned in Electronet® fences over night in areas that need concentrated fertilizing. These pens, comprising two or less rolls of fencing (depending on length) can be quickly moved from place to place. As the sheep settle for their afternoon rest period, shepherds could scout a new location and set up a new over-night pen in a matter of minutes. This type of fencing is so light (less than ten pounds per roll) that shepherds can easily carry rolls of it as they graze their flocks, but if a shepherd had more supplies to carry as well, he could use a donkey[44] as a pack animal. At night the donkey would double as a flock guardian.

Because shepherds' dogs are trained to patrol the boundaries of a grazing area, fencing would not be needed during the day. Two tending dogs can easily control hundreds of sheep, allowing the shepherd to merge several small flocks into one large flock. Because fencing would not be used, the need to cut paths for the fencing would also be eliminated and with it the frustration of planting fence posts in rocky soil. Thus, tending shepherds could pursue other activities like reading, painting or enjoying music as they oversaw their charges.

Jean Beck's experience in Orwell, Vermont is also worth studying from a tending perspective (even though she did not use dogs at all in her shepherding operation). In many ways her situation is similar to the clear blocks of British Columbia. Beck promised to keep the height of the grass in the town's cemetery at 2 1/2 inches and according to reports, her sheep succeeded in doing just that. But many of the town's people protested, not wanting their cemetery turned into a sheep pasture. As a result, Beck had to abandon the project.

I wonder, however, whether or not she could have reached a compromise with town officials if dogs had been an integral part of her operation. First, the sheep could graze for a few hours twice a day (early morning and evening) supervised by a shepherd. These hours would suit the natural grazing patterns of the sheep and would not offend public sensibilities. Second, if a shepherd were in attendance while the sheep grazed, he/she could talk to by passers and educate them about the project. Third, a shepherd could keep the flock from settling into one area. As a result, feces would not concentrate in any particular place. If sheep did defecate on grave markers, they could be quickly cleaned by the shepherd.

When the sheep were not grazing they could be placed in a small pen, either out of sight or off the cemetery grounds. In Beck's case, using Border Collies or other gathering dogs would have worked well because this particular cemetery was fenced. Not all cemeteries are, however, and where no fencing exists, tending dogs would save the shepherd the labor of constantly putting up and taking down Electronet®. In addition, the cemetery would not take on the appearance of a sheep pasture.

Tending methods would also work particularly well under the high voltage transmission lines in New Hampshire, even better in fact than on the ski slopes of Vermont. First, the shepherd in New Hampshire has only one flock of 500 very gregarious

---

44 The advantage to using a donkey is that it will eat the same fodder as a sheep.

Rambouillets instead of the three smaller flocks of 100 head (one group being semi-gregarious Montadales) used in Vermont. Second, as stated earlier, tending allows for controlled rotational or strip grazing without the use of fencing. Third, power line areas have natural boundaries that tending dogs can easily key off of, for example, the tree line on each side of the 400 foot wide area under the transmission lines. Moreover, the power line poles themselves can be boundaries defining a narrower graze area. Generally power line right-of-ways have dirt access roads which provide additional boundaries and as an added advantage, it is easier to transport sheep along dirt roads than to move them along grass paths.

Finally, when traveling long distances, extra people are not needed to cross roads and bypass properties. Instead, one shepherd and two tending dogs can easily control up to a thousand head of sheep. By patrolling along the length of a sheep column to keep them on a roadway, two tending dogs can protect private property by keeping the sheep from trampling it. And one dog can press the flock to the right side of the road, thus allowing traffic to pass by, while the other dog can work the opposite side of the flock, keeping them on the road.

In Canada's cut blocks and in the grassy Olympic Park in Calgary, and in the municipal parks of Edmonton and Fort Saskatchewan, sheep do graze without fences, but shepherds use Border Collies to keep them in their assigned areas. Although this does work, there are disadvantages inherent to the breed. Specifically, most Border Collies use "eye"[45] to control and move the livestock. This becomes a disadvantage when working in dense tall forage like clear blocks. When working in dense brush, sheep cannot easily see a Border Collie's eyes staring at them. Thus, they do not move away quite so easily.

This leaves Border Collies without their best tool to control livestock; they must now learn to bark or grip sheep. Although a skilled trainer can teach Border Collies to do this, it is not an easy task, for Border Collies have been bred to work quietly with little or no grip. The Bearded Collie, a dog with more bark and bite than the Border Collie, was developed in Great Britain for exactly this reason—to control livestock in dense forage areas (Combe 1987:169, Holmes 1984:61, 64).

Most Border Collies want their charges moving. Thus many dogs of this breed have difficulty allowing sheep to graze, and even if they are trained to do so, many become fixated on their charges and will not move to contain the flock unless prompted to do so by their handler. Finally, most Border Collies tend to flank wide, taking themselves away from the flock. This allows the sheep to trespass boundaries, either while grazing or moving along roadways.

---

45 "A most useful asset possessed by many of these border working collies was the power of "the eye", the ability to control the sheep by staring at them in a fixed steady manner. Dogs with the right amount of "eye" can keep their sheep bunched together well when driving them and thus avoid a great deal of flanking, running from one side to the other" (Grew: :1984:4)

A shepherd in Germany used Border Collies successfully while tending his flock. However, these dogs are bred with "loose eye," using grip to control their charges instead of eye and they work very close to the flock. They may be Border Collies in appearance and in heritage, but they work very much like German Shepherd Dogs.

Adventures in Tending

After returning from my first trip to Germany, I decided to incorporate tending techniques into my daily herding activities. We designed our farm to use the best abilities of Border Collies, but this resulted in certain problems. For example, when running sheep through handling equipment to vaccinate, worm or sort the flock, my Border Collies wanted to push sheep toward me from the rear of the flock. This often caused the flock to spread out and jam at the mouth of the chute. I needed the sheep flowing single file into the system. How could I accomplish this?

First, I thought that changing handling equipment was the answer. As I began looking through sheep supply catalogs, a heading caught my eye. It read: "Tired of 'Wrestling' with Your Sheep? Premier's Handling Equipment is Easier on You & Your Sheep!" The catalog was written in a unique style. It identified problems that shepherds often face and then suggested ways equipment could be used to solve these problems. For example the catalog said:

> …Sheep have four legs and can move on their own. That being so, design a system that "tricks" the sheep into moving by themselves when and where you want them to go, stopping them for treatment (vaccinating, drenching, branding, tagging, weighing, sorting, etc.) only when you put gates in their way. Effective sheep handling can be condensed into two words—CONTROLLED ESCAPE. Gates are carefully designed and arranged to easily meter and pack sheep into treatment chutes. Sheep are thus *self-caught* for treatment by panels on their sides and by fellow sheep or gates at front and rear.

It then asked:
"So how do I trick a sheep into doing what I want?"

and answered:
"The good news is that sheep behavior is predictable." Nearly all sheep:
- Fear water, mud, dogs, noise, and people (But in our case, they did not fear people).
- Move toward food, be it corn, hay or green pasture.
- Move toward lighted areas and away from dark areas when being driven.
- Move toward pen walls and gates that are "open" (allows sheep to see through) and away from pen walls & gates that are solid (sheep can't see through).

- Move away from confined, closed areas toward open areas.
- Move toward another sheep, especially ones they can follow.
- Have memories. They remember, prefer & follow familiar routes and gates.
- Prefer to cling to the outside walls of a pen when driven. They will follow a curved wall more readily than a straight one.
- Will move to pen corners & stand there. So if you want sheep to move, build pens with curved sides and with no 90° corners (Premier Sheep Supplies Catalog).

Thinking about this, I remembered from my trip that shepherds in Germany were able to "trick" their sheep into doing exactly these kinds of things without using expensive equipment. Each morning their flocks flowed smoothly out of holding pens to graze. Their unhurried exit was the response I was looking for. But how could I get my sheep to walk single file through a chute without jamming it?

To accomplish this I attached handling equipment to one of the gates of my tending pen (32 by 32 foot square) leaving the other as an entrance. I used a simple design: a "Y" chute, guillotine gate and head gate. But first I had to retrain my flock to pass through the system without fear because my Border Collies had made them associate the chute with fear of dogs. Thus, I needed to retrain them without using dogs.

To accomplish this I passed the sheep through the handling equipment by coaxing them into the chute with my voice and feeding them after they had exited. At the end of three days the sheep moved expeditiously through the system, showing no signs of fear. The true test, however, was to come the following weekend when I planned to drench (adversive) the sheep.

So, when that day arrived, I called to the sheep and they responded by entering the holding pen. Once the sheep were in the pen, I needed a dog but not one with Border Collie skills. Because Border Collies work from behind the flock they would jam the sheep at the mouth of the chute again. I needed a dog to help keep the mouth of the chute open so that one sheep could enter at a time. My first thought was to position a dog at the mouth of the pen, but I knew my Border Collies would stare at the sheep so intently they would be afraid to pass by. So I sent my German Shepherd, "Wren," into the pen. She did so by jumping in. Immediately the flock pressed against the outside wall adjoining the handling equipment. Meanwhile, I was standing in position to drench the sheep. Again I called out to the sheep and they freely moved toward the handling equipment. As the first few sheep entered the chute I called "Wren" to move up to a "pick" position at the mouth of the chute. The sheep continued to pass through the system single file. Once the chute was full, I began drenching the sheep. As the sheep exited the pen they were given a reward of grain to override the adversive effect of the drenching.

Although Wren kept steady pressure on the sheep in the holding pen as would a Border Collie, the response from the sheep was different. Because Wren was able to keep steady pressure on them from the mouth of the pen without using "eye" the sheep did not stress. As a result, they did not jam the chute.

Another problem I faced involved moving our sheep between their winter and summer pastures. These pastures were separated by a mile of roadway. To move between them the sheep had to cross both a two and four lane highway. Usually, I loaded them on a trailer with the help of my Border Collies and trucked them between pastures. Yet this way of doing things had a disadvantage; I needed to make several trips to transport the entire flock to the new pasture area.

Unfortunately, one day when I needed to move the sheep (70 head), the truck would not start, so instead, I had to use dogs. To do the job safely, though, I had to avoid many hazards while in route. Specifically, I had to keep the sheep from grazing in yards or flower beds, move the flock in such a way that traffic could pass by, and finally, take the sheep across two busy highways.

Before attempting the task, I talked with a friend who also worked with Border Collies. His situation was similar to mine; he also moved sheep along a road while in route to grazing areas, but he encountered a reoccurring problem. While his Border Collie could move the sheep along the road, it was not good at keeping the flock (300 head) out of the yards and flower beds.

Still, I didn't want to erect and take down fencing for it would take considerable time to do so. So I set off down the road with the sheep and my Border Collies. Unfortunately, traffic posed immediate problems. When a car approached (usually the

Photograph 12.1 "Y" chute added to a tending pen.

Photograph 12.2 Jean DeNapoli and Jetta running a flock of sheep through a handling system. Notice that the sheep are not jammed against the mouth of the chute. As the dog patrols back and forth the flock lines up to exit the pen. that bordered the road. He solved the problem by erecting portable Electronet® fence along the route.

driver did not want to wait for the flock to pass) the dogs would push the flock from the rear, thereby blocking the road way. If I flanked my dogs to push the flock over they would flank so wide (as Border Collies want to do) that they would run around the car. As a result, the sheep stayed in the middle of the road. This was unacceptable and I had to return the sheep to their winter pasture.

Then a friend suggested I use my German Shepherd, Wren, to deal with the problem. After all, I was training her to be a useful tending dog. I also convinced my friend to work her German Shepherd Dog (trained in the tending style) with mine (backed up by my husband and his Border Collie).

We each choose a side of the road to work. When we approached the highway, I held the flock at the junction by keeping Wren in front while my friend stopped traffic. Once it was safe to cross, I called to the flock and without incident, led them across the highway—so far, so good.

We then proceeded up a hill along a fenced field often full of sheep and cows. I prayed the animals would be in another field, for if not they would run to the fence, calling my flock to join them. I knew this would pose a serious challenge and I was not sure "Wren" could hold the flock on the roadway. Sure enough, those cattle and sheep were in the field bordering the road and as my flock heard the calls of the animals on the other side of the fence, they picked up their heads.

This was the warning sign; chaos would soon follow. So I asked my friend to leave the side of the flock she was working with her dog and move to the front. At least this would help control the speed at which chaos would occur and buy me some time. In any case, the Border Collies were not far away.

It worked! It bought me the time to press Wren into action. She felt the pressure of the flock to leave the road, but I wondered if she could hold them. The flock slowed down and "Wren" began patrolling the full length of the flock without a word from me. I called to the flock and fed the leader corn as we walked past the animals on the other side of the fence.

At the top of a hill about a half mile from our destination was an area filled with houses surrounded by beautifully landscaped yards. After we passed this area we would be home free. My friend moved her dog to the outside while I worked the inside. Now, just at this critical time, a car approached and neither dog had escorted a flock past a moving car before. To make matters worse, the dogs needed to switch sides, another exercise neither of us had practiced. We called our dogs and switched sides without incident. I sent Wren to press the sheep to the side of the road while my friend's dog kept them out of the flower gardens. The car moved slowly past the flock with just enough space to pass. Now we were almost at the new pasture, safe at last. After that, I took my flock along this same route many times without problems and I left my Border Collies at home for my tending dogs did just fine without their help.

Of course the tending method of herding sheep is not always, or by definition, the best method to use. Situations are often unique and each shepherd has to choose the kind of dog and the approach that works best given the uniqueness of that situation. Those who are unsatisfied for whatever reasons with conventional approaches, should think about tending. It might provide useful solutions to the problems you face.

Appendix I

Sheep Breeds

This is not a complete list. It is taken from *A World Dictionary of Breeds Types and Varieties of Livestock* written by I.L. Mason in 1951. Some sheep breeds are not listed if they numbered less than 10,000 animals in a given country. The names of the countries listed were known as such in 1951.

Austria
- Carinthian
- Merino
- Brownheaded Mutton
- Cross Breed

Belgium
- Campine
- Entre-Sambre-et-Meuse
- Suffolk
- Hampshire Down
- Texel

Bulgaria
- Bulgarian (Zackel type)
- Karnobat

Czechoslovakia
- Bohemia
- Moravia
- Merino
- Karakuls
- East Friesian
- Valachian Zackel
- Hampshire
- Tsigai

France
- *Arles Merino
- *German Improved Land
- *Précoce
- *Rambouillet

Merino ancestry

(Meat-wool breeds)
*Ile-de-France
Southdown
Berry
*Charmoise
Cotentin
Bluefaced Main
Avranchin
Texel
*Boulonnais
Caux

(Local Native Breeds North)
Sologne
Trun
Ushant

(Local Native Breeds Southwest-Pyrenean)
Central Pyrenean
Manech
Basque
Béarn
*Lourdes
Aure
Campan
Landes
Corbières

(Local Native Breeds South-Central)
Causses
Limousin
*Lacaune
Bizet
Velay Black
Rava
Ardes

(Local Native Breeds South-East)
Préalpes du Sud
Thônes
Marthod
Corsican

Germany
*Merino
*Mutton Merino
*German Improved Land
German Blackheaded Mutton
German Whiteheaded Mutton
East Friesian
Wilstermarsch
Heath
Leine
Skudde
Pomeranian
Bentheim
Rhön
German Mountain
Karakul
Hungary
*Hungarian Merinos
Tsigai
Racka (Zackel)

Italy
*Visso
*Upper Visso
*Improved Apulian
*Improved Lucanian
*Improved Calabrian
*Maremma

(Apennine type)
Garfagnana
Massa
Casentino
*Chiana
Siena
*Emilia Apennine
Frabosa
Perugian (hill and plain)

(Lop-eared alpine type)

Bergamo
Langhe
Biella
Lamon
Cadora
Varese
Alpago
Paduan
Pavullo
Friuli
Val Badia
Val Senales

(course wooled type)

Sardinian
Sicilian
Lecce
Altamura
Pagliarola
Campanian

Netherlands
- Texel
- Friesian

Poland
- Polish Heath
- Polish Zackel
- Swiniara
- *Précoce
- *Fagas
- Pomeranian

Portugal
- *Merino
- Bordaleiro
- Romania
- Turcana (Zackel)
- Tsigai
- Karakul
- Stogos
- Spanca
- Karnabat
- Racka

Spain
- *Merino
- *Mancha
- *Castilian
- *Aragon
- *Andalusian
- *Segura
- Lacho

Switzerland
- White Swiss Mountain
- Black-Brown Mountain
- Brown Headed MuttonValais

Yugoslavia
- Pramenka
- Tsigai
- Solcava

Persia
- Arabi
- Baluchi
- Farahani
- Gurgani
- Kalaku
- Kermani
- Khurasani
- Kurdi
- Luri
- Maku
- Turki
- Zandi

Iraq
- Awasi
- Arabi
- Kurdi

Syria, Palestine and Jordan
- Awasi
- Ak-Karaman
- Arabi

Egypt
- Abidi
- Ausimi
- Barki
- Fellahi
- Ibeidi
- Rahmani
- Saidi

Cyprus
- Cyprus Fat-tailed Awasi
- Ak-Karaman
- Arabi

Turkey
- Ak-Karaman
- Kizil-Karaman
- Dağlic
- Kivircik
- Karayaka
- Kamakyuruk
- Pirlak
- Chios
- Tushin
- *Turkish Merino

Appendix II

Plants Poisonous to Sheep

Alfalfa (Medicago sativa)
    Toxic parts: leaves and stems
    Toxic chemicals: coumestan, coumestrol and medicagenic acid
    General symptoms of poisoning:
        bloat
        erythema
        infertility
        skin, peeling of

Astragalus - A. lentiginosus
    Toxic parts: flowers, leaves, seeds. and stems
    Toxic chemicals: slaframine and swainsonine
    General symptoms of poisoning:
        abortion
        blisters, weeping
        brain, vacuolation of
        cytoplasm vacuolation
        death
        depression
        eyes, dull
        fetus, dead
        gait, unsteady
        heart rate, elevated
        incoordination
        kidney, congestion of

kidney, vacuolation of
lethargy
liver, congestion of
muscle, weakness of
sperm, detached tails
sperm mobility, poor
testicle degeneration
ventrical(right),edema
ventricle(right),round
weakness

Autumn crocus (Colchium autumnale)
Toxic parts : all parts
Toxic chemicals: colchiceine and colchicine
General symptoms of poisoning:
collapse
depression
diarrhea
salivation
Ingesting fresh leaves in the following quantities results in death (Cooper and Johnson 1984): 6.4 g/kg body weight lambs (2-3 months old)

Black cherry (Prunus serotina)
Toxic parts: leaves, seeds and twigs
Toxic chemicals: amygdalin and prunasin
General symptoms of poisoning:
breathing, labored
coma
convulsions
death by asphyxiation
gait, staggering
muscle spasms
paralysis
unconsciousness

Black locust (Robinia pseudoacacia)
Toxic parts: bark, leaves and seeds
Toxic chemicals: phasin and robin(in)
General symptoms of poisoning:
anorexia

Black oak (Quercus velutina)
Toxic parts: acorns and leaves
Toxic chemicals: gallic acid, pyrogallol and tannic acid
General symptoms of poisoning: N/A

Bog-laurel (Kalmia polifolia)
Toxic parts: all parts
Toxic chemicals: andromedotoxins
General symptoms of poisoning:
- depression
- gait, staggering
- nausea
- recumbency
- salivation
- vomiting

Bracken (Pteridium aquilinium)
Toxic parts: all parts
Toxic chemicals: aquilide A, prunasin, ptaquiloside and thiaminase
General symptoms of poisoning:
- blindness

Broom snakeweed (Gutierrezia sarothrae)
Toxic parts: leaves and stems
Toxic chemicals: alpha-pinene and gamma-humulene
General symptoms of poisoning:
- abortion
- anorexia
- constipation
- death
- diarrhea
- icterus
- weakness

Buckwheat (Fagopyrum esculentum)
Toxic parts: all parts
Toxic chemicals: fagopyrin
General symptoms of poisoning:
- blistering
- paralysis
- recumbency
- skin, peeling of

Bur buttercup (Ceratocephalus testiculatus)
Toxic parts: plant juices
Toxic chemicals: ranunculin
General symptoms of poisoning:
- anorexia
- breathing, labored
- death
- diarrhea
- dyspnea
- recumbency
- weakness

Castor bean (Ricinus communis)
Toxic parts: seeds
Toxic chemicals: ricin
General symptoms of poisoning: N/A
Experimental oral lethal doses for sheep is 1-2 g/kg

Chokecherry (Prunus virginiana)
Toxic parts: leaves, seeds and twigs
Toxic chemicals: amygdalin
General symptoms of poisoning:
- coma
- convulsions
- death by asphyxiation
- dyspnea
- paralysis

Climbing nightshade (Solanum dulcamara)
Toxic parts: immature fruit and leaves
Toxic chemicals: soladulcidine, solanine and solasodine
General symptoms of poisoning:
- death
- diarrhea
- falling down
- gait, staggering
- pupil dilation
- temperature, elevated

Common milkweed (Asclepias syriaca)
Toxic parts: latex, leaves and stems
Toxic chemicals: desglucosyrioside, syrioboside and syrioside
General symptoms of poisoning:
    death

Cut-leaved coneflower (Rudbeckia laciniata)
Toxic parts: flowers, leaves and stems
Toxic chemicals: unknown chemical
General symptoms of poisoning:
    anorexia
    depression
    incoordination

Cypress spurge (Euphorbia cyparissias)
Toxic parts: latex and seeds
Toxic chemicals: 5-deoxyingenol
General symptoms of poisoning: N/A

Death camas (Zigadenus venenosus)
Toxic parts: all parts
Toxic chemicals: zygacine
General symptoms of poisoning:
    breathing, shallow
    cyanosis
    death
    mouth, frothing of
    muscle, weakness of
    nasal discharge
    prostration
    salivation
    urination, frequent
    vomiting

Eastern whorled milkweed (Asclepias verticillata)
    Toxic parts: flower buds and leaves
    Toxic chemicals: galitoxin
    General symptoms of poisoning:
        agitation
        bloat
        breathing, labored
        convulsions
        death
        depression
        opisthotonos
        pupil dilation
        temperature, elevated
        trembling
        weakness, posterior

Entire-leaved groundsel (Senecio integerrimus)
    Toxic parts: all parts
    Toxic chemicals: unknown chemical

California rose-bay (Rhododendron macrophyllum)
    Toxic parts: leaves and stems
    Toxic chemicals: andromedotoxins
    General symptoms of poisoning
        ataxia
        colic
        depression
        recumbency
        vomiting
        weakness

Canadian milk-vetch (Astragalus canadensis)
    Toxic parts: flowers and leaves
    Toxic chemicals: 3-nitropropionic acid
    General symptoms of poisoning: N/A

Colorado rubberweed (Hymenoxys richardsonii)
    Toxic parts: all parts, leaves and stems
    Toxic chemicals: hymenovin
    General symptoms of poisoning:
        death
        kidney, congestion of
        liver, congestion of
        lungs, congestion of
        vomiting

English Ivy (Hedera helix)
    Toxic parts: leaves, mature fruit and plant juices
    Toxic chemicals: didehydrofalcarinol, falcarinol and hederasaponins
    General symptoms of poisoning: N/A

European spindletree (Euonymous europaeus)
    Toxic parts: bark, leaves and seeds
    Toxic chemicals: evomonoside and evonine
    General symptoms of poisoning: N/A

False hellebore (Veratrum viride)
    Toxic parts: rhizome, roots and young shoots
    Toxic chemicals: germidine and jervine
    General symptoms of poisoning:
        nausea
        salivation

Field horsetail (Equisetum arvense)
    Toxic parts: leaves, seeds and stems
    Toxic chemicals: thiaminase
    General symptoms of poisoning: N/A

Five-hooked bassia (Bassia hyssopifolia)
    Toxic parts: flowers and leaves
    Toxic chemicals: oxalate
    General symptoms of poisoning:
        coma
        death
        gait, unsteady
        incoordination
        tetany
        weakness

Foxglove (Digitalis purpurea)
    Toxic parts: all parts
    Toxic chemicals: digitoxin
    General symptoms of poisoning; N/A

Garden-sorrel (Rumex acetosa)
    Toxic parts: leaves and stems
    Toxic chemicals: oxalate
    General symptoms of poisoning:
        coma
        death
        incoordination
        mouth, frothing of
        pupil dilation
        recumbency

Greasewood (Sarcobatus vermiculatus)
    Toxic parts: leaves
    Toxic chemicals: oxalate
    General symptoms of poisoning:
        coma
        death
        depression
        prostration
        weakness

Iceland Poppy (Papaver nudicaule)
    Toxic parts: all parts and plant juices
    Toxic chemicals: physiological alkaloids (unknown chemical)
    General symptoms of poisoning:
        bloat
        incoordination
        muscle spasms
        nervousness
        recumbency

Jimsonweed (Datura stramonium)
Toxic parts: all parts
Toxic chemicals: atropine, hyoscine (scopolamine) and hyoscyamine
General symptoms of poisoning:
- ataxia
- breathing, rapid
- collapse
- dyspnea
- gait, unsteady
- incoordination
- recumbency
- reflex excitability
- trembling
- water intake, reduced

Kentucky coffeetree (Gymnocladus dioicus)
Toxic parts: leaves and seeds
Toxic chemicals: cytisine
General symptoms of poisoning:
- death

Lamb's-quarters (Chenopodium album)
Toxic parts: leaves and stems
Toxic chemicals: nitrate and oxalate
General symptoms of poisoning: N/A

Leafy spurge (Euphorbia esula)
Toxic parts: latex
Toxic chemicals: 5-deoxyingenol
General symptoms of poisoning:
- blistering
- death

Locoweed (Oxytropis sericea)
Toxic parts: flowers, leaves, mature fruit, seeds and stems
Toxic chemicals: swainsonine and slaframine
General symptoms of poisoning:
- abortion
- carpal joint, flexure
- coat, rough and dry
- death
- depression
- eyes, dull
- incoordination
- nervousness
- recumbency

Low larkspur (Delphinium bicolor)
Toxic parts: leaves and seeds
Toxic chemicals: methyllycaconitine
General symptoms of poisoning: N/A

Marsh arrow-grass (Triglochin palustris)
Toxic parts: flowers and leaves
Toxic chemicals: triglochinin
General symptoms of poisoning:
- convulsions
- nervousness
- recumbency
- salivation

Marsh horsetail (Equisetum palustre)
Toxic parts: leaves and stems
Toxic chemicals: palustrine and thiaminase
General symptoms of poisoning:
- diarrhea
- muscle, weakness of
- sweating

Menzies larkspur (Delphinium menziesii)
Toxic parts: all parts
Toxic chemicals: methyllycaconitine
General symptoms of poisoning: N/A

Appendix II: Plants Poisonous to Sheep

Naked-flowered sneezeweed (Helenium flexuosum)
    Toxic parts: leaves and stems
    Toxic chemicals: flexuosin A and flexuosin B
    General symptoms of poisoning:
        convulsions
        dyspnea
        weakness

Oleander (Nerium oleander)
    Toxic parts: all parts
    Toxic chemicals: oleandrin
    General symptoms of poisoning:
        breathing, rapid
        death
        mouth, irritation of

Cultivated onion (Allium cepa)
    Toxic parts: bulbs and leaves
    Toxic chemicals: N-propyl disulphide, SMCO and oxalate
    General symptoms of poisoning:
        anemia

Poison suckleya (Suckleya suckleyana)
    Toxic parts: leaves and stems
    Toxic chemicals: unknown cyanogenic glycoside
    General symptoms of poisoning:
        collapse
        death by asphyxiation
        dyspnea
        heart rate, slow
        salivation

Poison-hemlock (Conium maculatum)
Toxic parts: flowers, leaves, mature fruit, roots, seeds, stems and young shoots
Toxic chemicals: coniine, gamma-coniceine and N-methylconiine
General symptoms of poisoning:
- ataxia
- carpal joint, flexure
- death
- defecation, frequent
- salivation
- tail, kinked
- trembling
- urination, frequent
- weakness

Pokeweed (Phytolacca americana)
Toxic parts: all parts
Toxic chemicals: phytolaccigenin
General symptoms of poisoning:
- diarrhea
- drowsiness
- gastroenteritis
- vomiting

Potato (Solanum tuberosum)
Toxic parts: immature fruit, leaves, stems and tubers
Toxic chemicals: chaconine
General symptoms of poisoning:
- death
- incoordination
- weakness

Purple locoweed (Oxytropis lambertii)
Toxic parts: flowers, leaves and mature fruit
Toxic chemicals: swainsonine
General symptoms of poisoning:
- abortion
- agitation
- carpal joint, flexure
- death
- incoordination

Appendix II: Plants Poisonous to Sheep

Red oak (Quercus rubra)
Toxic parts: acorns, immature fruit and leaves
Toxic chemicals: gallic acid, pyrogallol and tannic acid
General symptoms of poisoning: N/A

Reed canarygrass (Phalaris arundinacea)
Toxic parts: leaves
Toxic chemicals: gramine, hordenine and 5MMethyltryptamine
General symptoms of poisoning:
    coma

Scarlet pimpernel (Anagallis arvensis)
Toxic parts: all parts
Toxic chemicals: cyclamin
General symptoms of poisoning: N/A

Seaside arrow-grass (Triglochin maritima)
Toxic parts: all parts, flowers and leaves
Toxic chemicals: triglochinin and taxiphillin
General symptoms of poisoning:
    convulsions
    death by asphyxiation
    nervousness
    recumbency
    salivation
    trembling
    vomiting

Sheep sorrel (Rumex acetosella)
Toxic parts: leaves and stems
Toxic chemicals: oxalate
General symptoms of poisoning:
    coma
    death
    falling down
    gait, staggering
    muscle spasms
    nasal discharge

Sheep-laurel (Kalmia angustifolia)
    Toxic parts: all parts
    Toxic chemicals: andromedotoxins
    General symptoms of poisoning:
        ataxia
        coma
        convulsions
        death
        depression
        dyspnea
        headache
        nasal discharge
        pupil dilation
        recumbency
        salivation
        vomiting

Showy milkweed (Asclepias speciosa)
    Toxic parts: leaves, mature fruit and seeds
    Toxic chemicals: desglucosyrioside and syrioside
    General symptoms of poisoning:
        appetite, loss of
        breathing with grunts
        breathing, labored
        recumbency

Silky lupine (Lupinus sericeus)
    Toxic parts: leaves, mature fruit, seeds and stems
    Toxic chemicals: anagyrine, sparteine and lupanine
    General symptoms of poisoning:
        breathing, labored
        coma
        convulsions
        death by asphyxiation
        depression
        dyspnea

Silvery lupine (Lupinus argenteus)
    Toxic parts: leaves and seeds
    Toxic chemicals: lupanine and anagyrine
    General symptoms of poisoning:
        breathing, labored
        coma
        convulsions
        death by asphyxiation
        trembling

Small lupine (Lupine pusillus)
    Toxic parts: leaves and seeds
    Toxic chemicals: unknown
    General symptoms of poisoning:
        breathing, labored
        coma
        convulsions
        death by asphyxiation
        trembling

Spotted water-hemlock (Cicuta maculata)
    Toxic parts: all parts
    Toxic chemicals: cicutol
    General symptoms of poisoning:
        breathing, labored
        death
        recumbency

St. John`s-wort (Hypericum perforatum)
    Toxic parts: flowers and leaves
    Toxic chemicals: hypericin
    General symptoms of poisoning:
        convulsions
        erythema
        skin, peeling of

Sudan grass (Sorghum sudanense)
Toxic parts: leaves and stems
Toxic chemicals: dhurrin and nitrate
General symptoms of poisoning:
- breathing, labored
- erythema
- itchiness
- nasal discharge

Sun spurge (Euphorbia helioscopia)
Toxic parts: latex
Toxic chemicals: 12-deoxyphorbol
General symptoms of poisoning:
- abdominal pains
- coma
- death
- diarrhea
- lungs, congestion of
- mouth, irritation of
- salivation
- vomiting

Tall larkspur (Delphinium glaucum)
Toxic parts: all parts
Toxic chemicals: methyllycaconitine
General symptoms of poisoning: N/A

Tansy ragwort (Senecio jacobaea)
Toxic parts: all parts
Toxic chemicals: jacobine and seneciphylline
General symptoms of poisoning: N/A

Timber milk-vetch (Astragalus miser)
Toxic parts: leaves and stems
Toxic chemicals: miserotoxin and 3-nitropropanol
General symptoms of poisoning:
- breathing, labored
- cyanosis
- death
- incoordination

Two-grooved milk-vetch (Astragalus bisulcatus)
Toxic parts: flowers, leaves and stems
Toxic chemicals: selenium and swainsonine
General symptoms of poisoning:
- ascites
- brain, vacuolation of
- coat, rough and dry
- cytoplasm vacuolation
- depression
- fetus, dead
- kidney, vacuolation of

Velvety goldenrod (Solidago mollis)
Toxic parts: leaves and stems
Toxic chemical: unknown chemical
General symptoms of poisoning:
- breathing, rapid
- death
- nausea
- vomiting

Western minniebush (Menziesia ferruginea)
Toxic parts: leaves
Toxic chemicals: andromedotoxins
General symptoms of poisoning:
- breathing, labored
- gait, staggering
- mouth, frothing of
- nausea
- paralysis
- salivation
- weakness

Western water-hemlock (Cicuta douglasii)
Toxic parts: leaves, roots and young shoots
Toxic chemicals: cicutoxin
General symptoms of poisoning:
- bloat
- coma
- convulsions
- death
- death by asphyxiation
- gait, unsteady
- incoordination
- lesions, no specific
- mouth, frothing of
- muscle spasms
- muscle twitching
- nervousness
- salivation
- tarsal joint knuckling
- teeth grinding
- trembling
- urination, frequent

White camas (Zigadenus elegans)
Toxic parts: all parts
Toxic chemicals: zygacine
General symptoms of poisoning:
- ataxia
- breathing, rapid
- coma
- death
- mouth, frothing of
- nasal discharge
- nausea
- salivation
- urination, frequent
- vomiting

White rose-bay (Rhododendron albiflorum)
    Toxic parts: leaves
    Toxic chemicals: andromedotoxins
    General symptoms of poisoning:
        convulsions
        death
        nasal discharge
        paralysis
        salivation

White snakeroot (Eupatorium rugosum)
    Toxic parts: leaves and stems
    Toxic chemicals: tremetol
    General symptoms of poisoning:
        acidosis
        death
        nervousness
        trembling

Wild cabbage (Brassica oleracea L.) includes common cultivated crops such as kale, broccoli, Brussels sprouts, and cabbage.
    Toxic parts: all parts, flowers and leaves
    Toxic chemicals: glucosinolates and SMCO
    General symptoms of poisoning:
        Heinz bodies
        hemoglobinuria

Wild radish (Raphanus raphanistrum)
    Toxic parts: seeds
    Toxic chemicals: glucosinolates
    General symptoms of poisoning: N/A

Yellow sage (Lantana camara)
    Toxic parts: immature fruit and leaves
    Toxic chemicals: lantadene A & B
    General symptoms of poisoning: N/A

Appendix III

Dog Breeds

SHEEP DOGS

ARMANT	EGYPT
BEAUCERON	FRANCE
BEDOUIN SHEPHERD DOG	ISRAEL
BELGIAN SHEPHERD	BELGIUM
BERGAMASCO	ITALY
BERGER DE PICARD	FRANCE
BRIARD	FRANCE
CANAAN DOG	ISRAEL
CATALAN SHEEPDOG	SPAIN
CHEVENNES SHEPHERD	FRANCE
CROATIAN SHEEPDOG	YUGOSLAVIA
DUTCH SHEPHERD DOG	NETHERLANDS
EAST EUROPEAN SHEPHERD	RUSSIA
GERMAN SHEEPPOODLE	GERMANY
GERMAN SHEPHERD DOG	GERMANY
ICELAND DOG	ICELAND
MUDI	HUNGARY
POLISH LOWLAND SHEEPDOG	POLAND
PORTUGUESE SHEEPDOG	PORTUGAL
PULI	HUNGARY
PYRENEAN SHEPHERD	FRANCE
SCHAPENDOES	NETHERLANDS
STEPPENOWTCHARKA	RUSSIA
TIBETAN TERRIER	TIBET
YUGOSLAVIAN SHEEPDOG	YUGOSLAVIA

Appendix III: Dog Breeds

REINDEER DOGS

FINNISH LAPPHUND	FINLAND
LAPINPOROKOIRA	FINLAND
SAMOYED	RUSSIA
SWEDISH LAPPHUND	SWEDEN

CATTLE DOGS

APPENZELL MOUNTAIN DOG	SWITZERLAND
BERNESE MOUNTAIN DOG	SWITZERLAND
BOUVIER DES ARDENNES	BELGIUM
BOUVIER DES FLANDRES	BELGIUM
ENTELBUCH MOUNTAIN DOG	SWITZERLAND
GIANT SCHNAUZER	GERMANY
GREATER SWISS MOUNTAIN DOG	SWITZERLAND
PORTUGUESE CATTLE DOG	PORTUGAL
PUMI	HUNGARY
ROTTWEILER	GERMANY
STANDARD SCHNAUZER	GERMANY
SWEDISH VALLHUND	SWEDEN

GUARD DOGS

AIDI	MOROCCO
AKBASH DOG	TURKEY
ANATOLIAN	TURKEY
CASTRO LABOREIRO	PORTUGAL
CAUCASIAN OWTCHARKA	RUSSIA
ESTRELA MOUNTAIN DOG	PORTUGAL
GREEK SHEEPDOG	GREECE
HOVAWART	GERMANY
ILLYRIAN SHEEPDOG	YUGOSLAVIA
ISTRIAN SHEEPDOG	YUGOSLAVIA
KARABASH	TURKEY
KOMONDOR	HUNGARY
KUVASZ	HUNGARY
MAREMMA SHEEPDOG	ITALY
MIDDLE ASIAN OWTCHARKA	RUSSIA
OWCZAREK PODHALANSKI	POLAND
PERRO DE PASTOR MALLORQUIN	BALEARIC ISLANDS, SPAIN
PYRENEAN MASTIFF	SPAIN
PYRENEAN MOUNTAIN DOG	RANCE
RAFEIRO DO ALENTEJO	PORTUGAL

RUMANIAN SHEEPDOG	ROMANIA
SLOVAK CUVAC	CZECHOSLOVAKIA
SOUTH-RUSSIAN OWTCHARKA	RUSSIA
SPANISH MASTIFF	SPAIN
TATRA MOUNTAIN SHEEPDOG	POLAND
TIBETAN MASTIFF	TIBET

Breed Descriptions—Pictures of these dogs can be found in:
- *Eyewitness Handbook of Dogs* by David Alderton Published by Dorling Kindersley, New York 1993.
- The Continental Kennel Club All-Breed Handbook—1-800-952-3376.

Armant

Height: 56 cm (22 inches)
Weight: 50-65 lbs (110 kg to 143 kg)
Coat: long, rough and shaggy
Color: black, black/tan, gray, grayish yellow

Beauceron

Height: 61 to 70 cm (24 to 27 1/2 in.)
Weight: 66-85 lbs (30-38 1/2 kg.)
Coat: short dense
Color: black with tan markings: below the eyes, muzzle, chest, throat, feet, under the tail. Also available in black, or harlequin

Belgian Sheepdog

Height: 56 to 61 cm (22 to 24 in.) female; 61 to 66 cm (24 to 26 in.) male.
Weight: 62-66 pounds (28-30 kg.)
Coat: Groenendael and Tervuren are long-haired varieties. Malinois is short haired. Rough-haired Laekenois is rough coated.
Color: Groenendael is black may have small white markings as well. Tervuren and Malinois, shades of red, fawn or grey with black overlay. Laekenois: reddish fawn with black overlay.

Bergamasco

Height: 56 to 61 cm (22 to 24 in.)
Weight: 26 to 38 kg (57 to 84 lb)
Coat: heavily corded
Color: shades of grey including salt and pepper

Berger Picard

Height: 55 to 60 cm (21 1/2 to 23 1/2 in.) females; 60 to 65 cm (21 1/2 to 25 1/2 in.) males.
Weight: 50-70 pounds (23-31 kg.)
Coat: medium, long rough, straight heavy undercoat
Color: shades of fawn to grey.

Briard

Height: 53.3 to 63.5 cm (21 to 25 in.) female; 58.4 to 68.6 cm (23 to 27 in.) male.
Coat: coarse, long, dry and goatlike.
Color: all solid colors except white.

Canaan Dog

Height: 19-24 inches (48 to 60 cm)
Weight: 35-56 lbs. (16 to 25 kg)
Coat: medium short, harsh and straight, plumed tail
Color: white with large patches of brown, black or red can also be brown or black without white

Catalan Sheepdog

Height: 18- 20 1/2 inches (45 to 52 cm)
Weight: 35-45 lbs. (16 to 20 kg)
Coat: long and wavy
Color: black, black/tan, fawn with black tips, grizzle, or brindle

Croatian Sheepdog

Height: 16-21 inches (40 to 53 cm)
Weight: 30-48 lbs. (13.5 kg to 22 kg)
Coat: 3-6 inches long soft but dense and wavy, shorter on head and front legs
Color: black with some white

Dutch Shepherd

(Same as Belgian Shepherds except for color three varieties: Long, Short, and Rough haired)
Coat: Long, Short and Rough haired
Color: various shades of brindle (gold, red, yellow, silver and gray)

East European Shepherd

Height: 24-29 inches (60 to 73 cm)
Weight: 75-113 lbs. (34 kg to 51 kg)
Coat: short and smooth
Color: black, black/tan, sable, seldom white or brindle

German Sheeppoodle

Height: 19-24 inches (48 to 61 cm)
Weight: 59-65 lbs. (26 to 29 kg)
Coat: long shaggy
Color: white, roan, pied

German Shepherd

Height: 22-26 inches (56-66 cm)
Weight: 75-95 pounds (34-43 kg)
Coat: rough, long rough, long
Color: black with tan markings, light grey with black and dark saddle. May have small white markings on the chest or inside of legs. Solid black or white.

Mudi

Height: 14-20 inches (35-50 cm)
Weight: 18-29 lbs. (8-13 kg)
Coat: long 2 inches on body, short on head, course
Color: black, white, or brindle

Polish Lowland Sheepdog

Height: 40-52 cm (16 to 20 in.)
Weight: 13.5 to 16 kg (30 to 35 lb)
Coat: long, thick, shaggy
Color: any color
Tail: missing or docked

Portuguese Sheepdog

Height: 16-22 inches (40 to 55cm)
Weight: 25-45 lbs. (11 to 20 kg)
Coat: long and wavy, course, shaggy on head
Color: yellow, brown ,fawn, black, grey

Puli

Height: 40.6 to 45.7 cm (16 to 18 in.) male; 36.8 to 40.6 cm (14 1/2 to 16 in.) female.
Weight: 13 to 15 kg (28 1/2 to 33 lb) male; 10 to 13 kg (22 to 28 1/2 lb) female
Coat: long, coarse corded
Color: black, reddish black, grey, white and apricot

Pyrenean Shepherd Dog

Height: 40 to 48 cm (15 3/4 to 19 in.) male; 38 to 46 cm (15 to 18 in.) female
Weight: 8 to 15 kg (17 1/2 to 33 lbs.)
Coat: long shaggy, corded, goat like
Color: tawny, solid or with black, may have white on chest and feet; harlequin of various shades

Schapendoes

Height: 43 to 51 cm (17 to 20 in.)
Weight: 15 kg (33 lb)
Coat: long dense
Color: all colors. Generally blue-grey to black.

Tibetan Terrier

Height: 14-16 inches (35 to 40 cm)
Weight: 16-32 lbs. (7 to 14 kg)
Coat: long shaggy
Color: white, black, gray, gold, with or without white

Appendix IV

Breed Contacts

Kennel Clubs
American Kennel Club
260 Madison Avenue
New York, NY 10016
website: akc.org

Canadian Kennel Club
200 Ronson Drive, Suite 400
Etobicoke, Ontario M9W 5Z9
website: ckc.ca

United Kennel Club, Inc.
100 East Kilgore Road
Kalamazoo, MI 49001
website: ukcdogs.com

The Kennel Club
1 Clarges Street
London, England W1Y 8AB
website: thekennelclub.org.uk

Continental Kennel Club
P.O. Box 908
Walker, LA 70785
website: continentalkennelclub.com

Fédération Cynologique Internationale
Place Albert 1er, 13
B-6530 Thuin
Belgique
website: fci.be

Appendix V

Puppy Aptitude Test Sheet

Hear My Voice: An Old World Approach to Herding

SCORING THE RESULTS

Following are the responses you will see and the score assigned to each particular response. You will see some variations and will have to make a judgment on what score to give them.

	TEST	PURPOSE	SCORE	
SOCIAL ATTRACTION	Place puppy in test area. From a few feet away the tester coaxes the pup to her / him and kneeling down. Tester must coax in a direction away from the point where it entered the testing area.	Degree of social attraction, confidence, or dependence	Came readily, tail up, jumped, bit at hands	1
			Came readily, tail up, pawed, licked at hands	2
			Came readily, tail up	3
			Came readily, tail down	4
			Came hesitantly, tail down	5
			Didn't come at all	6
FOLLOWING	Stand up and walk away from the pup in a normal manner. Make sure the pup sees you walk away.	Degree of following indicates independence.	Followed readily, tail up, got underfoot, bit at feet	1
			Followed readily, tail up, got underfoot	2
			Followed readily, tail up	3
			Followed readily, tail down	4
			Followed hesitantly, tail down	5
			Did not follow or went away	6
RESTRAINT	Crouch Down and gently roll the pup on its back and hold it with one hand for a full 30 seconds.	Degree of dominante or submissive tendency. How it accepts stress when social / physically dominated.	Struggled fiercely, flailed, bit	1
			Struggled fiercely, flailed	2
			Settled, struggled, settled with some eye contact	3
			Struggled, then settled	4
			No struggle	5
			No struggle, strained to avoid eye contact	6
SOCIAL DOMINANCE	Let the pup stand up and gently stroke him from the head to the back while you crouch beside him. Continue stroking until a recognizable behavior is established.	Degree of acceptance of social dominance. Pup may dominate by jumping and nipping, or is independent and walks away	Jumped, pawed, bit, growled	1
			Jumped, pawed	2
			Cuddled up to tester and tried to lick face	3
			Squirmed, licked at hands	4
			Rolled over, licked at hands	5
			Went away and stayed away	6
ELEVATION DOMINANCE	Bend over and cradle the pup under its belly, fingers interlaced, palms up, and elevated just off the ground. Hold it there for 30 seconds	Degree of accepting dominance while in position of no control	Struggled fiercely, tried to bite	1
			Struggled fiercely	2
			Struggled, settled, struggled, settled	3
			No struggle, relaxed	4
			No struggle, body stiff	5
			No struggle, froze	6
RETRIEVING	Crouch beside pup and attract his attention with crumbled up paper ball. When the pup shows interest and is watching, toss the object 4-6 feet infront of the pup.	Degree of willingness to work with a human. High coloration between ability to retrieve and successful guide dogs, obedience dogs, and field trial dogs.	Chased object, picked it up and ran away	1
			Chased object, stood over it and did not return	2
			Chased object, picked it up and returned with it to tester	3
			Chased object and returned without it to tester	4
			Started to chase object, lost interest	5
			Does not chase object	6
TOUCH SENSITIVITY	Take the pups webbing of one front foot and press between finger and thumb lightly then more firmly until you get a response, while you count slowly to 10. Stop as soon as the pup pulls away or shows discomfort.	Degree of sensitivity to touch.	8-10 count before response	1
			6-8 count before response	2
			5-6 count before response	3
			3-5 count before response	4
			2-3 count before response	5
			1-2 count before response	6
SOUND SENSITIVITY	Place pup in the center of area test her or assistant makes a sharp noise a few feet from the pup. A large metal spoon struck sharply on a large metal pan works well.	Degree of sensitivity to sound. (Also can be a rudimentary test for deafness)	Listened, located sound and ran toward it barking	1
			Listened, located sound and walked slowly toward it	2
			Listened, located sound and showed curiosity	3
			Listened and located sound	4
			Cringed, backed off and hid behind tester	5
			Ignored sound and showed no curiosity	6
SIGHT SENSITIVITY	Place pup in center of room. Tie string around a towel and jerk it a few feet away from the puppy.	Degree of intelligent response to strange object.	Looked, attacked and bit object	1
			Looked and put feet on object and put mouth on it	2
			Looked with curiosity and attempted to investigate, tail up	3
			Looked with curiosity, tail down	4
			Ran away or hid behind tester	5
			Hid behind tester	6
STABILITY	An umbrella is opened about five feet from the puppy and gently placed on the ground.	Degree of startle response to a strange object	Looked and ran to the umbrella, mouthing or biting it	1
			Looked and walked to the umbrella, smelling it cautiously	2
			Looked and went to investigate	3
			Sat and looked, but did not move toward the umbrella	4
			Showed little or no interest	5
			Ran away from the umbrella	6

Wendy Volhard's Puppy Aptitude Test © 1981, 2000, 2005 Used by permission.

Bibliography

Alderton, David. Dogs: *The Visual Guide to Over 300 Dog Breeds from Around the World*. New York: Dorling Kindersley, 1993.

Allix, André. "The Geography of Fairs: Illustrated by Old-World Examples." *The Geographical Review* 12 (1922): 532-69.

Anderson, D. M., and C. V. Hulet, W. L. Shupe, J. N. Smith, and L. W. Murray. "Response of Bonded and Non-bonded Sheep to the Approach of a Trained Border Collie." *Applied Animal Behaviour Science* 21 (1988): 251-57.

Arnold, G. W. and R. A. Maller, "An Analysis of Factors Influencing Spatial Distribution in Flocks of Grazing Sheep." *Applied Animal Behaviour Science* 14 (1985): 173-89.

Atwell, Gerry. "Wolf Predation on Calf Moose." *Journal of Mammalology* 45 (1964): 313-14.

Backhaus, Dieter. "Experimentelle Untersuchungen über die Sehschärfe und das Farbsehen einiger Huftiere." *Zeitschrift für Tierpsychologie* 16 (1959) 445-67.

Baker, Don. *The Way of the Shepherd*. Portland, Ore.: Multnomah, 1987.

Bartlett, Melissa. "A Novice Looks at Puppy Aptitude Testing." *Pure Bred Dogs, AKC Gazette* (March 1979): 31-42.

_____. "Puppy Aptitude Testing." *Pure Bred Dogs, AKC Gazette* (March 1985): 31-34, 64.

Bauman, Diane L. *Beyond Basic Dog Training*. New York: Howell Book House, Inc., 1986.

Behan, Kevin. *Natural Dog Training: The Canine Arts Kennel Program, Teach Your Dog by Using His Natural Instincts*. New York: William Morrow and Co., 1992.

Billingham, Viv. *One Woman and Her Dog*. Moffat, Scotland: Lochar, 1984.

Blackford, R. H., Jr. "Multispecies Systems for California." In *Proceedings, Conference on Multispecies Grazing, June 25-28, 1985*. Edited by Frank H. Baker and R. Katherine Jones. Morrilton, Ark.: Winrock International Institute for Agricultural Development, 1985.

Booth, Sheila and Gottfried Dildei. *Schutzhund Obedience Training in Drive*. Ridgefield, Conn.: Podium, 1992.

Bremner, K. J., and J. B. Braggins and R. Kilgour. "Training Sheep as 'Leaders' in Abattoirs and Farm Sheep Yards." *Proceedings of the New Zealand Society of Animal Production* 40 (1980): 111-16.

Broadbent, Kate. "Grazing." *Ranch Dog Trainer* 13 (1998): 16-17.

Bueler, Lois E. *Wild Dogs of the World*. New York: Stein and Day, 1973.

Busch, Robert H. *The Wolf Almanac*. New York: Lyons and Burford, 1995.

Caplin, Linda and Suzanne Clothier. *Agility Training Workbook: A Step-by-Step Motivational Approach*. Frenchtown, N.J.: Flying Dog Press, 1989.

Carrier, E. H. *Water and Grass: A Study in the Pastoral Economy of Southern Europe*. London: Christophers, 1932; reprint New York: AMS Press, 1980.

Childs, Nathan B. *Leader of the Pack: Shaping Dog Instincts through Pack Training*. Panama, Fla.: Pack, 1986.

Ciucci, Paolo and Luigi Boitani. *Viability Assessment of the Italian Wolf and Guidelines for the Management of the Wild and a Captive Population*. Bologna [Italy]: Instituto nazionale di biologia della selvaggina "Alessandro Ghigi," 1991.

Combe, Iris. *Herding Dogs: Their Origins and Development in Britain*. London and Boston: Farber and Farber, 1987.

Coppinger, Lorna and Raymond Coppinger. "Livestock-guarding Dogs that Wear Sheep's Clothing." *Smithsonian* (1982): 64-73.

Corbett, Lawrence K. *The Dingo in Australia and Asia*. Ithaca, N.Y.: Comstock/Cornell University Press, 1995.

Coren, Stanley, *The Intelligence of Dogs*. New York: Free Press, 1994.

Crabbe, Barb and John Lyons. "Stop Kick." *Horse and Rider* 33 (1995): 86-89.

Daley, Michael. "A Day on the Slopes with Joshua Moody, Shepherd." *Sheep Magazine* (June 1994): 16.

D'Ambrosio, Anthony, Jr. and Steven D. Price, *Schooling to Show Basics of Hunter-Jumper Training*. New York: Viking, 1978.

Davies, Elwyn. "The Patterns of Transhumance in Europe." *Geography* 26 (1941): 155-68.

Davidar, E. R. C. "Ecology and Behavior of the Dhole or Indian Wild Dog Cuon alpines (Pallas)." In *The Wild Canids: Their Systematics, Behavioral Ecology, and Evolution*. Edited by Michael W. Fox. New York: Van Norstrand Reinhold, 1975.

De Prisco, Andrew and James B. Johnson. *Mini-Atlas of Dog Breeds*. Neptune City, N.J.: T. F. H. Publications, 1990.

Downs, James F. *Animal Husbandry in Navajo Society and Culture*. Berkeley: University of California Press, 1964.

Dunbar, Ian. *Dog Behavior: Why Dogs Do What They Do*. Neptune, N.J.: T. F. H. Publications, 1979.

El Aich, A. and L. R. Rittenhouse. "Herding and Forage Ingestion by Sheep." *Applied Animal Behaviour Science* 19 (1988): 279-90.

Elliott, Rachel Page. *The New Dogsteps: A Better Understanding of Dog Gait through Cineradiography ("Moving X-rays")*. New York: Howell Book House, 1983.

Evans, E. E. "Transhumance in Europe." *Geography* 25 (1940): 172-80.

Ferster, C. B. and B. F. Skinner. *Schedules of Reinforcement*. New York: Appleton-Century-Crofts, 1957.

Fisher, Gail Tamases and Wendy Volhard. "Puppy Personality Profile." *Pure Bred Dogs, AKC Gazette* (March 1985): 36-42.

Fogt, Bruce. "How Do I See What I Saw?" *The Working Border Collie* (January/February 1992): 6.

"For Hire: Sheep Labor." *Success* (June 1993): 10.

Fox, Michael W. *The Dog: Its Domestication and Behavior*. New York and London: Garland STPM Press, 1978.

Geist, Valerius. *Mountain Sheep: A Study in Behavior and Evolution*. Chicago: University of Chicago Press, 1971.

Gentry, Christine. *When Dogs Run Wild: The Sociology of Feral Dogs and Wildlife*. Jefferson, N.C.: McFarland, 1983.

Gerhard, Laura K. "Rent a Sheep, Save a Forest." *National Wool Grower* (September 1993): 18.

Gilbert, Edward M., Jr. and Thelma R. Brown. *K-9 Structure and Terminology*. New York: Howell Book House, 1995.

Grigg, David B. *The Agricultural Systems of the World: An Evolutionary Approach*. London: Cambridge University Press, 1974.

_____. *The Dynamics of Agricultural Change: The Historical Experience*. New York: St. Martin's Press, 1982.

Gómez-Ibáñes, Daniel Alexander. *The Western Pyrenees: Differential Evolution of the French and Spanish Borderland*. Oxford: Clarendon, 1975.

Grew, Sheila. *Key Dogs from the Border Collie Family*. Sudbury, Suffolk: Payn Essex, 1984.

Hancock, John. "Studies in Monozygotic Cattle Twins IV, Uniformity Trials: Grazing Behavior." *New Zealand Journal of Science and Technology* 2 (32) (1950): 22-59.

Henson, Elizabeth. *British Sheep Breeds*. Princes Risborough: Shire, 1986.

Holland, Vergil S. *Herding Dogs: Progressive Training*. New York: Howell Book House, 1994.

Holmes, John. *The Farmer's Dog*. London: Popular Dogs, 1984.

_____. *The Family Dog: Its Choice and Training*. 10th ed. London: Popular Dogs, 1994.

The Holy Bible, New International Version. Grand Rapids: Zondervan, 1978.

Hutson, G. D. "The Influence of Barley Food Rewards on Sheep Movement through a Handling System." *Applied Animal Behaviour Science* 14 (1985): 263-73.

_____. "A Note on the Preference of Sheep for Whole or Crushed Grains and Seeds." *Animal Production* 38 (1984): 145-46.

Irigaray, Louis and Theodore Taylor. *A Shepherd Watches, A Shepherd Sings*. Garden City, N.Y.: Doubleday, 1977.

Jones, Arthur F. and Ferelith Hamilton. *The World Encyclopedia of Dogs*. New York: Galahad, 1971.

Jones, H. Gwyn and Barbara C. Collins. *A Way of Life: Sheepdog Training, Handling and Trialling*. Ipswich: Farming Press, 1987.

Keller, W. Phillip. *A Shepherd Looks at Psalm 23*. Grand Rapids: Zondervan, 1970.

_____. *A Shepherd Looks at the Good Shepherd and His Sheep*. Grand Rapids: Zondervan, 1978.

_____. *A Child's Look at the Twenty-third Psalm*. New York: Bantam Doubleday, 1981.

_____. *Lessons from a Sheepdog*. Dallas: Word, 1983.

Kendrick, Keith. "Through a Sheep's Eye." *New Scientist* 126 (May 12, 1990): 62-65.

Koehler, William R. *The Koehler Method of Dog Training*. New York: Howell Book House, 1962.

Kruesi, William. *The Sheep Raiser's Manual*. Charlotte, Vt.: Williamson, 1985.

Lantinga, J. H. *Schapen Houden als Liefhebberij*. Zutphen: Thieme, 1976.

Lawrence, A. B. and D. G. M. Wood-Gush. "Social Behaviour of Hill Sheep: More to it than Meets the Eye." *Applied Animal Behaviour Science* 17 (1987): 382.

Lenburg, Jeff. *The Encyclopedia of Animated Cartoons*. New York: Facts on File, 1991.

"Let the Sheep Do the Shearing." *Mother Earth News*. (August–September, 1991): 14.

Lichtner-Hoyer, Peter. *Complete Cavalletti*. New York: Breakthrough, 1991.

Lithgow, Scott. *Training and Working Dogs for Quiet Confident Control of Stock*. St. Lucia, Qld.: University of Queensland Press, 1987. [See also Appendix by Don Morris.]

Lopez, Barry Holstun. *Of Wolves and Men*. New York: Scribner, 1978.

Love, Myron. "Alberta continues Rent-a-Sheep Programs to Mow Grass." *Sheep Magazine* (September 1994): 5.

"Lunch on the Lawn." *Maclean's* (June 29, 1992): 6.

Lynch, J. J., G. N. Hinch, and D. B. Adams. *The Behaviour of Sheep: Biological Principles and Implications for Production*. Wallingford: C. A. B. International, 1992.

Lyon, McDowell. *The Dog in Action: A Study of Anatomy and Locomotion as Applying to All Breeds*. New York: Howell Book House, 1974.

Mason, Ian L. *A World Dictionary of Breeds, Types and Varieties of Livestock*. Slough, Bucks.: Commonwealth Agricultural Bureaux, 1969.

McLeroth, Diane. *The Briard: A Collection*. N.p.: Aubry Associates and The Briard Club of America, 1982.

Mech, L. David. "Hunting Behavior in Two Similar Species of Social Canids." In *The Wild Canids: Their Systematics, Behavioral Ecology, and Evolution*. Edited by Michael W. Fox. New York: Van Norstrand Reinhold, 1975.

_____. *The Wolf: The Ecology and Behavior of an Endangered Species*. Garden City, N.Y.: Published for The American Museum of Natural History by the Natural History Press, 1970.

_____. *The Way of the Wolf*. Stillwater, Minn.: Voyageur, 1991.

Meyer, F. B. *The Shepherd Psalm*. Grand Rapids: Kregel, 1991.

Milani, Myrna M. *The Body Language and Emotion of Dogs*. New York: Morrow, 1986.

Monks of New Skete. *The Art of Raising a Puppy*. Boston: Little, Brown and Co., 1991.

_____. *How to Be Your Dog's Best Friend: A Training Manual for Dog Owners*. Boston: Little, Brown, and Co., 1978.

Mowatt, Farley. *Never Cry Wolf*. Boston: Little, Brown, and Co., 1963.

Munro, Derek B. "Canadian Poisonous Plants." Canadian Poisonous Plants Information System. http://www.cbif.gc.ca/pls/pp/poison?p_x=px.

Murray, Sidman. *Coercion and Its Fallout*. Boston: Authors Cooperative, 1989.

Nation, Allen. "Cowboy Heaven." *The Stockman Grass Farmer* 48 (September 1991): 1.

O'Donnell, Dick. "Mary had a Little Lamb . . ." *Sheep Magazine* 9 (October–November 1988): 16.

Ollivant, Alfred. *Bob, Son of Battle.* New York: Grosset and Dunlap, 1898.

Outerbridge, David. *The Last Shepherds.* New York: Viking, 1979.

Paulsen, Gary. *The Haymeadow.* New York: Dell, 1992.

Pecoult, Alain. "Sheepdog News from France." *Working Sheepdog News* 3 (September 1980): 26-28.

Pfaffenberger, Clarence. *The New Knowledge of Dog Behavior.* New York: Howell Book House, 1963.

Pryor, Karen. *A Dog and a Dolphin.* North Bend, Wash.: Sunshine, 1995.

_____. *Don't Shoot the Dog: The New Art of Teaching and Training.* New York: Bantam, 1984.

_____. *Lads before the Winds: Diary of a Dolphin Trainer.* Third ed. North Bend, Wash.: Sunshine, 1994.

_____. *On Behavior: Essays and Research.* North Bend, Wash.: Sunshine, 1995.

The Reader's Digest Illustrated Book of Dogs. Pleasantville, N.Y.: Reader's Digest Association, 1989.

Rennie, Neil. *Working Dogs: Breeding, Feeding, Training and Care.* Auckland: Moa Beckett, 1992.

Ruzzo, Patty. "Four Steps to a Better Ring Performance." Workshop handout, Colchester, Conn., 1990.

_____. "Food for Thought." Workshop handout, Colchester, Conn., 1990.

Ryder, Michael L. *Sheep and Man.* London: Duckworth, 1983.

_____. "Sheep." In *Evolution of Domesticated Animals.* Edited by Ian L. Mason. London and New York: Longman, 1984.

Saenger, Paul F. "Economics of Short Duration Grazing: Sheep. Handout, University of Vermont Extension Service, Burlington, 1984.

Sárkány, Pál and Imre Ócsag. *Hungarian Dog Breeds.* Budapest: Corvina Kiadó, 1986.

Schaller, George B. *Mountain Monarchs: Wild Sheep and Goats of the Himalaya.* Chicago: University of Chicago Press, 1977.

Schwantje, H. M. *Interim Guidelines for the Use of Domestic Sheep for Vegetation Management in British Columbia.* Revised ed. [Victoria]: Province of British Columbia, Ministry of Forests, 1995.

Seitz, A. "Untersuchungen über das Formsehen und optische Grössenunterscheidung bei der Skudde (ostpreussisches Landschaft)." *Zeitschrift für Tierpsychologie* 8 (1951): 424-41.

Self, Robert. *Dogs Self-Trained: Basic and Advanced Training Manuals.* Galesburg, Ill.: R. T. and Associates, 1981.

Sheep Production Handbook. Denver: Sheep Industry Development Program, 1988.

"Sheep and Trees Mix Well." *The Stockman Grass Farmer* (February 1991): 14.

Simmons, Paula. *Raising Sheep the Modern Way*. Charlotte, Vt.: Garden Way, 1989.

Skinner, B. F. *Contingencies of Reinforcement: A Theoretical Analysis*. New York: Appleton-Century-Crofts, 1969.

Slemming, C. W. *He Leadeth Me*. Fort Washington, Pa.: Christian Literature Crusade, 1942.

Smythe, R. H. *The Dog: Structure and Movement*. New York: Arco, 1970.

Spira, Harold R. *Canine Terminology*. New York: Howell Book House, 1982.

Starr, Douglas. "A Sheep's Best Friend." *National Wildlife* 24 (June 1986): 14-19.

Steinhart, Peter. *The Company of Wolves*. New York: Knopf, 1995.

Stephanitz, Max von. *The German Shepherd in Word and Picture*. Jena: Anton Kämpfe, 1925.

Syme, L. A. "Social Disruption and Forced Movement Orders in Sheep." *Animal Behaviour* (1981): 283-88.

Thomas, Elizabeth Marshall. "Man's Next-Best Friend." *New York Review of Books* (1995): 42.

United States Department of Agriculture, Soil Conservation Service. "Benefits of Short Duration Grazing," n.d.

University of Vermont Extension Service and American Agriculturalist Magazine. "Voisin Pasture Management System." Franklin County Demonstration and Research Project, 1984.

"Using Goats for Brush Control in Humid Climates." *The Stockman Grass Farmer* (February 1991): 7.

Vett, Mark A. G. "Behaviour of New Zealand Sheepdogs: Their Mature Behaviour, Ontogeny, and Evolution." MSc thesis, University of Auckland, 1982.

Voisin, André. *Grass Productivity*. Washington, D.C.: Island Press, 1988.

Volhard, Joachim J. and Gail Tamases Fisher. *Training Your Dog: The Step-by-Step Manual*. New York: Howell Book House, 1983.

Warren, Jerry T. and Ivar Mysterud. "Summer Habitat Use and Activity Patterns of Domestic Sheep on Coniferous Forest Range in Southern Norway." *Journal of Range Management* 44 (January 1991): 2-6.

"Woolly Weed-eaters." *Canadian Geographic* (May–June 1992): 10.

Videography

Pryor, Karen and Gary Wilkes. *Shaping: Building Behavior with Positive Reinforcement*. VHS. North Bend, Tex.: Sunshine, 1992.

_____. *Sit! Clap! Furbish! How to Teach Cues and Establish Behavioral Control*. VHS. North Bend, Tex.: Sunshine, 1994.

"Schutzhund with Gottfried Dildei: Advanced Obedience." DVD. Littleton, Co.: Canine Training Systems.

"Training Sheep Herding Dogs with Karl Fuller." DVD. Menomonie, Wis.: Leerburg.

Wilkes, Gary. *Click! and Treat Training Kit*. Version 1.1. VHS. [Phoenix]: Video Media, 1996.

Other training videos are available at Canine Training Systems (www.caninetrainingsystems.com) and at Leerburg (leerburg.com).

www.ingramcontent.com/pod-product-compliance
Lightning Source LLC
Chambersburg PA
CBHW041241240426
43668CB00025B/2456